Chinese firms and technology in the reform era

China has undergone rapid economic change since the introduction of reforms in the early 1980s. The country has become more open to technological and economic influences from outside.

In this study Yizheng Shi analyses the technological behaviour of state owned firms. In particular he shows how they have imported, utilised and assimilated foreign technology into their operations. The author argues that despite being granted more autonomy and having to face increased competition, Chinese firms are still not motivated to properly assimilate imported technology because of the absence of well delineated property rights.

With a strong analytical framework supported by empirical evidence, **Yizheng Shi** provides fresh insights on China's rapidly developing economy.

Yizheng Shi is an Assistant Professor of Chinese Business Studies at the School of Business, Hong Kong Baptist University. His research interests include international technology transfer, foreign direct investment and technology policy in China.

Routledge Studies in the Growth Economies of Asia

1 **The Changing Capital Markets of East Asia**
Edited by Ky Cao

2 **Financial Reform in China**
Edited by On Kit Tam

3 **Women and Industrialization in Asia**
Edited by Susan Horton

4 **Japan's Trade Policy**
Action or Reaction?
Yumiko Mikanagi

5 **The Japanese Election System**
Three Analytical Perspectives
Junichiro Wada

6 **The Economics of the Latecomers**
Catching-Up, Technology Transfer and Institutions in Germany, Japan and South Korea
Jang-Sup Shin

7 **Industrialization in Malaysia**
Import Substitution and Infant Industry Performance
Rokiah Alavi

8 **Economic Development in Twentieth Century East Asia**
The International Context
Edited by Aiko Ikeo

9 **The Politics of Economic Development in Indonesia**
Contending Perspectives
Edited by Ian Chalmers and Vedi Hadiz

10 **Studies in the Economic History of the Pacific Rim**
Edited by Sally M. Miller and A. J. H. Latham

11 **Workers and the State in New Order Indonesia**
Vedi R. Hadiz

12 **The Japanese Foreign Exchange Market**
Beate Reszat

13 **Exchange Rate Policies in Emerging Asian Countries**
Edited by Stefan Collignon, Jean Pisani-Ferry and Yung Chul Park

14 **Chinese Firms and Technology in the Reform Era**
Yizheng Shi

Chinese firms and technology in the reform era

Yizheng Shi

ROUTLEDGE
London and New York

First published 1998
by Routledge
11 New Fetter Lane, London EC4P 4EE

Simultaneously published in the USA and Canada
by Routledge
29 West 35th Street, New York, NY 10001

Typset in Times by RefineCatch Limited, Bungay, Suffolk
Printed and bound in Great Britain by Biddles Ltd, Guildford and King's Lynn

British Library Cataloguing in Publication Data
A catalogue record for this book is available from the British Library

Library of Congress Cataloging in Publication Data
Shi, Yizheng, 1948–
Chinese firms and technology in the reform era / Yizheng Shi.
p. cm.
Includes bibliographical references and index.
1. Government business enterprises – Effect of technological innovations on – China. 2. Technology transfer – China – Case studies.
I. Title.
HD3861.C6S54 1998
338′.064′0951 – dc21 97–27161
CIP

ISBN 0–415–17141–5

Contents

List of tables and figures

LIST OF TABLES

LIST OF FIGURES

Acknowledgements

This book is the result of four years' research. It would never have been completed without the encouragement and active support of a number of people for whom my gratitude is substantial.

I am deeply grateful to Dr Paul Robertson who provided me with the most needed encouragement and substantive academic advice. He patiently read various drafts of every chapter of the manuscript, and made many thoughtful and important comments. I am also indebted to Professor Wolfgang Kasper, Associate Professor Paul McGavin and Dr On Kit Tam for their important suggestions and detailed comments.

I gratefully acknowledge financial assistance from the School of Economics and Management, the University College, the University of New South Wales to support this study and allow me to carry out field research in China.

Finally, I would like to express my heartfelt gratitude to my wife Xian Li who has taken the responsibility of maintaining a family so that I could spend more time on the research. Without her great understanding, encouragement and self-sacrifice, the research never could have been completed.

Abbreviations

BEI	The Bureau of Electronics Industry (of local governments)
CAIA	The China Audio Industry Association
CKD	complete knocked down
CMRS	The Contract Management Responsibility System
EPC	The Economic and Planning Commissions (at provincial and local levels in China)
FDI	foreign direct investment
GBCBI	The General Bureau of Communications and Broadcasting Industry (of the Ministry of Machine Building and Electronics Industry)
GVIO	gross value of industrial output
ICRS	Internal Contract Responsibility System
MoFERT	The Ministry of Foreign Economic Relations and Trade (of China)
MoMBEI	The Ministry of Machine Building and Electronics Industry (of China)
MRS	Manager Responsibility System
NIE	newly industrialized economy
RMB	renminbi (the Chinese currency)
RMRB	Renmin Ribao (The People's Daily)
SEC	The State Economic Commission (of China)
SETC	The State Economic and Trade Commission (of China)
SEZ	special economic zones
SOE	state owned enterprises
SPC	The State Planning Commission (of China)
SSB	The State Statistic Bureau (of China)
TD	tape deck
TR	tape recorder
TRI	tape recorder industry
TVE	township and village enterprises

1 Introduction

THE QUESTIONS TO BE TACKLED

International technology transfer has played a very important role in the economic development of the world. Throughout history, new ideas and technical knowledge have spread from one region to another through various channels and led to productivity increases and economic growth. Many nations, for example, the United States and Japan, have greatly benefited from imported technological knowledge for their rapid industrialization and economic growth. In the past thirty years, the economic success of the newly industrialized economies (NIEs) has once more proved the importance of international technology transfer by means of openness and integration into the global economy (World Bank 1991: 88).

Historical evidence shows that the critical problem for a recipient country of foreign technology is how to get imported technology to work effectively. Many economists have pointed out that recipients must engage in deliberate and intensive local learning programmes to assimilate technology if technology transfer is to be successful. (Nelson 1990a; Rosenberg and Frischtak 1985; and Stewart 1990). Nevertheless, there have been controversial arguments regarding such issues as to where the local learning efforts should come from and how to guide these efforts in technology transfer (Stewart 1990; Cimoli and Dosi 1990). Some people think that the market mechanism fails in dealing with transactions of international technology transfer to developing countries. Therefore, government interventions in the process are necessary (Wade 1990). Others (for example, Ernst and O'Connor 1989; Nelson 1990b) argue that properly motivated individual firms guided by correct signals from the market are the key to effective technology transfer.

This book is going to tackle the issue of effective technology transfer by investigating the technology transfer process of Chinese state owned enterprises (SOEs) during the reform era. It will discuss the impact of institutional arrangements on the technological behaviour of SOEs in transferring foreign technology. Specifically, it looks into the actions and motivations of Chinese SOEs in the tape recorder industry in importing and utilizing foreign technology during the period between 1980 and 1993.

Traditionally, political power in China was centralized and the state controlled the nation through a bureaucratic hierarchy. Many people argue that it is this institutional arrangement which prevented modern technology from originating in China (Needham 1969; Bodde 1991; and Wang 1993). After the establishment of the People's Republic, the state mobilized enormous resources to industrialize the economy with imported technology. Nevertheless, China was still technologically backward and productivity growth was very slow before the economic reforms (Chen *et al.* 1988). The aim of economic reforms started in the late 1970s is to increase the economic efficiency of the whole economy and of industrial enterprises in particular. To these ends, advanced foreign technology and equipment have been imported and utilized on an unprecedentedly large scale by Chinese enterprises.

In the early 1980s, China changed her industrialization policy from an 'extensive' growth strategy which relied on establishing new plants, to an 'intensive' one involving the technological renovation of existing enterprises.[1] As a result, one of the major changes in China's technology import policy was a shift of emphasis from the construction of a small number of large turn-key projects, which was the normal practice of the 1960s and the 1970s, to the importation of key pieces of equipment and technologies to update the production facilities of existing industrial enterprises, especially those of small- and medium-sized enterprises.[2] It was argued by Chinese policy makers that this approach would promote 'technological progress' in existing enterprises and improve the effectiveness and efficiency of technology imports by Chinese enterprises (Sheng *et al.* eds 1991). As a result of the policy change, imports of technology to renovate existing enterprises became the most important form of technology transfer to China. It has been reported that during the period 1979 to 1988, Chinese state owned enterprises (SOEs) undertook about 20,000 import projects which cost more than US $1.5 billion (Sheng *et al.* eds 1991: 81).

The change in China's technology import policy after 1980 was in line with and a part of the overall economic reform programme and the 'open-door' policy.[3] Because of economic reforms and the 'open door' policy, many fundamental aspects of economic life in China changed dramatically. In the case of the operation of SOEs, economic decisions were decentralized to a considerable extent and enterprises obtained enlarged decision making autonomy in their daily operation. Furthermore, market forces were gradually introduced to replace central planning and the administrative distribution of industrial inputs and products, and SOEs were made responsible for their own businesses by linking their rewards to their performance through various responsibility systems (Tidrick and Chen Jiyuan 1987; World Bank 1990; Wu Jinglian 1993a).

Changes in the institutional setting and market environment certainly affected the behaviour of SOEs in general, and their technological behaviour in importing and utilizing foreign technology in particular. The purpose of this study is to look into how institutional changes resulting from economic reforms affected the technological behaviour of SOEs in transferring foreign technology. Specifically, this study will investigate a number of Chinese SOEs in the tape recorder industry that carried out technology import projects during the 1980s under the programme of renovating existing enterprises with imported technology. This study tries to find out: (1) how SOEs performed in carrying out their technology import projects within the programme of renovating existing enterprises with imported technology, and in utilizing the imported technology afterwards; and (2) how changes in institutional arrangements as the result of economic reforms affected the technological behaviour of SOEs in transferring foreign technology.

In the remaining part of this chapter, technological development in traditional China and the institutional legacies of traditional Chinese society will be discussed to provide a historical and societal background for the discussion in the following chapters. Although this book looks into the technological behaviour of state owned enterprises in the 1980s, an understanding of the institutional tradition of China is necessary. This is because the impact of the traditional bureaucratic system and central control can still be felt in the operation of state owned enterprises in general and their technological behaviour in particular. Finally, this chapter defines the scope of the study and briefly describes the major content of each chapter of the book.

TECHNOLOGICAL DEVELOPMENT IN TRADITIONAL CHINESE SOCIETY

Backwardness in science and technology is often given as one of the major reasons why China is poor and underdeveloped in the contemporary time. However, China has really only lagged behind in the advance of science and technology over the last several hundred years, probably since the fifteenth century. If we take Chinese history as a whole, the achievements of China in science and technology stood far in front of other civilizations for quite a long time. As documented in Needham's monumental works of *Sciences and Civilization in China* (1954–88), many important inventions that affected human history originated in China. Examples include the magnetic compass, paper, printing and gun powder, which are called proudly by Chinese scholars and in China's primary school textbooks as the great four inventions (Xu and Fan 1982). Furthermore, many scholars convincingly argue that as far as the flow of scientific and technological knowledge between West and East is concerned, Europe was primarily a recipient until the sixteenth century (Needham 1969; Volti 1982; and Sivin 1980).

Nevertheless, the numerous important inventions in China remained in a primitive form with very little refinement throughout history. Generally speaking, technical aspects of these inventions were purely in the form of personal know-how of illiterate artisans and, therefore, were not systematically summarized and recorded, due to the gap between scholars and artisans (Bodde 1991: 360). Furthermore, the many techniques were normally either in the area of public civil engineering, to keep the centralized empire running, or for the manufacturing of luxury goods for the emperor and government officials. They were seldom used widely for private profits and thus the efficiency of social production did not increase. In other words, economic growth of traditional China was mainly dependent on the supply of production factors of labour and land, rather than on technological progress. Consequently, economic growth in traditional China was not self-sustained (Rostow 1970: Chapter 1).

Eventually, the development of technology in traditional China, which was slow but steady in the past, stagnated in the later centuries. In sharp contrast with China, Europe made great advances in science and technology and overtook China around the late fifteenth century (Needham *et al.* 1969). After the Industrial Revolution, China fell even further behind. In spite of the glorious achievements in the past, China ultimately failed to develop modern science and technology.

Many scholars (for example, Needham 1969; Rostow 1970; Rosenberg and Birdzell 1986) argue that the reason for the stagnation of technology and science in traditional China since the late fifteenth century is deeply rooted in the social, economic, political and cultural institutions of traditional Chinese society. The development of Chinese civilization was less connected with other civilizations. After the ancient feudal period, China became a centralized empire as early as 221 BC. Since then, a special form of society developed, which is called Asiatic bureaucracism (Needham 1969) or officialism (Balazs 1964). This type of society is very different from the West, as the emperor had absolute authority over his subjects and ruled the country through a bureaucracy system. Some scholars (for example, Needham 1969; Bodde 1991; and Wang 1993) think that this system is much more rational than the social system in medieval Europe and contributed to the initial superiority of China in technology. Nevertheless, becoming increasingly rigid later on, this system lost its merit and turned out to be the obstacle to further development.

In imperial China, the emperor, who claimed to be the son of heaven, owned the whole country and collected taxes. The property of his subjects was not guaranteed by any form of law. It was quite common for the wealth of powerful people to be confiscated, when they lost the trust of the emperor and the empire was in financial trouble. As a result, people who had resources were reluctant to invest their money in any sort of capital goods, except land (Balazs 1964: 53; Jones 1987: Chapter 5; and Rosenberg and Birdzell 1986: 120).

The legitimacy of the emperor came from Confucianism, which was only one of the so-called 'one hundred schools of learning' before it became established as the orthodoxy of imperial China around 2,000 years ago (Xu and Fan 1982: 24–6). In Confucianism, the family is the basic unit constituting the society. The state is organized as a huge family and ruled by the emperor as if he is an authoritarian father (Bodde 1991: 194–203).

The emperor rules his country through a bureaucracy system which is staffed with non-hereditary elite selected through strict imperial examinations. People are selected to be civil servants, that is mandarins, based on their ability rather than their birth. Therefore, in a certain sense, this system encourages people to learn and draws the most intelligent people into the ruling class.

Since the whole country is owned by the emperor, the only avenue open to people who seek power, privilege and wealth is to be the servants of the emperor selected through the strict imperial examinations. The examination is very difficult. Prospective candidates need to spend much time and enormous efforts to prepare for it. Once a person passes the examination, he is awarded an entry ticket to a position in the bureaucracy system, and he immediately becomes a member of the scholar-officials. He enjoys outstanding social status and rules the ordinary people in his jurisdiction as their parent-official. Since the only thing he needs to do is to satisfy his superiors and there is no control from below, he can easily get rich through corruption to compensate for several years of study. As a Chinese proverb says, a person can acquire everything through reading classical books, i.e. preparing himself for the examinations.

Thus, scholar-officials, or mandarins, were the highest among the four major classes in the traditional Chinese society. They enjoyed every privilege, held all the power and thus were quite wealthy. Since people could only join this class through learning, this process was important to stabilize the social order of imperial China. This is because the system of imperial examinations attracted people with the best brains and what they had to learn to pass the examinations were basically the classics of Confucian philosophy which formed the ideological base of the empire.

According to the Confucian philosophy, it is the natural law of society that scholars rule those who labour with their brawn. Therefore, the role of scholars is to govern, to handle people and to play political games. What they learn are the classics of sages in Confucian school, which have nothing to do with science and technology. They, by nature, are against any kind of specialization. Therefore, the unproductive scholar-officials are not specialists but bureaucrats. To pass the examinations, mandarins have to concentrate their efforts on these basically humanistic studies. They need to remember and stick to all these classics and are not allowed to branch out into other lines of inquiry. Consequently, scholar-officials of traditional China were quite ignorant of other more specialized and scientific subjects (Bodde 1991: 213–19; Balazs 1964: 15–18).

The economic base of traditional China was a huge number of peasants. They submitted taxes in physical forms of grain and fabrics to maintain the bureaucracy system. They were forced to provide labour service to build public civil engineering projects. When needed, they had to provide military service as soldiers of the imperial army. Although they were entitled to be the second highest social class in imperial China, because they were the base of the state, peasants were bound to the soil by the government through a neighbourhood administration system, in which the large landlords, who normally were the clan elders of large local families, were responsible for collection of taxes and fulfilment of labour services. These numerous rural communities constituted traditional Chinese society which remained agrarian throughout the imperial period. The techniques used by the peasants, though highly developed, were traditional and have little changed over the centuries. The rural communities, which were the basic cells of the empire, were self-sufficient. Therefore, there was little need for them to exchange with each other (Elvin 1973).

Since the empire consisted of numerous self-sufficient agricultural communities, it is not surprising that the state wished to keep this system stable and restrict the activities of artisans and merchants. Artisans were at the lower level of society. They included craftsmen, engineers, architects, and skilled and unskilled labour other than peasants. Like the peasants, the mobility of artisans was restricted and their production activities were controlled by the government through a registry system. They had to produce certain products and submit them to the government. Since Confucianism, by nature, looks down upon physical labour and technical skills, manual workers were separated from mental workers. Consequently, artisans were normally illiterate and prevented from becoming rich with their technical skills, although the major inventions and technological achievements of ancient China have been made by them (Bodde 1991: 212–30).

Many scholars (for example, Needham 1969; Bodde 1991; Balazs 1964) attribute the failure of China to develop modern science and technology to the fact that merchants and industrialists were held back by the government and an independent civil class in the European sense has not emerged. From the view of the imperial government, merchants and industrialists were potentially disturbing factors to the state; they competed with the state in exploiting peasants and artisans. Thus, they could ruin the base of the state, and so their activities were regarded as immoral and had to be restricted (Bodde 1991: 217–21; Needham 1969: 197).

Consequently, in imperial China, merchants were discriminated against. They could be rich. But being wealthy in itself did not allow merchants to have higher social status. To be gentlemen, who were respected and powerful, people had to pass the imperial examinations and enter the bureaucracy class. Further, the property of merchants was not guaranteed by law. They also had to bribe their greedy parent-officials for protection. As a result, when merchants managed to accumulate wealth, their ambitions were to become a member of the bureaucracy by investing their money in land and in educating their children rather than in expanding their production (Balazs 1964: 32). Under the institution of scholar-officialism, acquisition of wealth and social status through Schumpeterian entrepreneurship was prohibited (Baumol 1990: 901).

To sum up, the reason that technological development stagnated in the later imperial China is due to the institutional arrangements which favoured redistributive rent-seeking rather than productive innovation. The incomes of social production were distributed among the members of traditional Chinese society according to social positions determined by the examination system which had the ability to reinforce the institutions of scholar-official bureaucracy. Therefore, the efforts and entrepreneurship of people were directed at getting the special social status, and resources were not invested in the application of innovation in industry and commerce activities. The final result of this game was negative and traditional China inevitably lost its dynamics in technological progress and economic growth (Baumol 1990; Buchnan 1980; and Jones 1988).

The institutional tradition of centralized political control through a bureaucratic hierarchy was so deep-rooted in the society that it survived into the modern era. In 1949, the People's Republic of China under the control of the Communist party was established. Unlike the previous governments, from the beginning, the Communist state emphasized science and technology and was committed to industrialization. To achieve the goal of rapid industrialization, and because of its ideology, the government consolidated state control over economic activities and established a central planning system throughout the country (Barnett 1981; Volti 1982; Riskin 1987; and Wang 1993). As a result, the traditional bureaucratic system was properly integrated with the socialist regime in the early 1950s, although the communist ideology is very different from Confucianism. Under the modern version of the bureaucratic system, all resources were owned by the state and distributed among the members of the society administratively. Consequently, there was still no incentive for Schumpeterian innovation and technological development in China was very slow before the reform, although the state invested enormous resources into the science and technology sector.

THE BOOK'S OUTLINE

This book is organized around the general theme of the technological behaviour of Chinese SOEs in importing and utilizing foreign technology during the reform era. In particular, it focuses on the change in the technological behaviour of SOEs as the institutional arrangements governing their operation changed. Therefore, this study follows the whole process of technology transfer in enterprises, as it has proceeded alongside the process of economic reforms in the past fifteen years.

The focus of this study is on the technological behaviour of enterprises. Thus, the study is at the firm level and investigates the behaviour of individual enterprises. Among the macroeconomic factors this study concentrates on those which have direct and significant impact on the technological behaviour of enterprises. Consequently, reforms in the state enterprise governance system, the government's policy on technological change and technology imports, and competition in the commodity markets which the specific enterprises served are highlighted.

Furthermore, this study is not an analysis of the general operation and performance of enterprises under state ownership. It is principally a study of how enterprises make decisions regarding their technological activities and how their technological activities are performed under certain conditions.

In this study, technology is broadly defined to include capital goods and disembodied technology as well as firm-specific and accumulated experience.[4] Furthermore, the technological behaviour of enterprises in transferring foreign technology is examined in depth from a number of different dimensions. These angles include the technological activities that the sample enterprises undertook, the objectives of enterprises in undertaking these activities, how they performed in undertaking these activities, and what factors affected their decisions and performance in undertaking these activities.

This book consists of eight chapters. Chapter 2 lays the theoretical foundations for the study and discusses some important issues about methodology. It starts with a discussion of the nature of technology and technological change. In the light of the theory of technological change, it argues that improvements in the technological capability of recipient firms can only be achieved through active learning. The chapter then discusses the role of institutional arrangements in economic life and proposes that a property rights approach be adopted to analyse the behaviour of SOEs. It argues that macroeconomic factors concerning institutional arrangements are fundamental in defining the technological behaviour of enterprises, and discusses the institutional reasons for the technological behaviour of SOEs. Finally, based on the theories discussed, an analytical framework is formulated to direct the empirical investigation.

The last part of Chapter 2 concentrates on the issue of research method. It argues that due to the nature of the research topic, a qualitative method is appropriate and realistic. Then it moves on to a discussion of how to measure the technological behaviour of enterprises and their performance in transferring foreign technology. In order to collect information on the whole process of technology transfer in enterprises, the technique of semi-structured questionnaire interviews and detailed case studies is used. Finally, it reports the actual process of field research, in which a number of enterprises in the Chinese tape recorder industry were selected as the sample firms at which questionnaire interviews were conducted. After a discussion of the criteria for the selection of sample firms, it reports what kinds of information were collected.

Chapter 3 presents the environmental background to this study. This chapter consists of two parts. The first talks about economic reforms in China after the late 1970s and analyses the impact of economic reforms on SOEs. In line with the focus of this study, the discussion centres on the reform of state enterprises. After a discussion of the major characteristics of state enterprises during the pre-reform period, the main aspects of the economic reforms are briefly discussed: macroeconomic reforms, state enterprises reforms and the development of the non-state sector. Then, the chapter looks at the changed behaviour of state enterprises during the different stages of economic reforms. It argues that due to the blurred property right arrangements, SOEs tended to have a short term perspective, which affected their approach to long term technological investment.

The second part of Chapter 3 discusses technology imports in China during the 1980s. It starts with a review of technology imports before the reform. It then discusses changes in technology and technology import policy, with a focus on the programme of renovating existing enterprises with imported technology. Finally, it describes the technology import practices of Chinese enterprises during the 1980s and discusses the main shortcomings of enterprises in importing and utilizing foreign technology.

Chapters 4 and 5 analyse the empirical information collected by the general survey of the sample enterprises in the Chinese tape recorder industry. Chapter 4 concentrates on technology import projects carried out by the sample firms and their performance in utilizing the imported technology. It first presents an account of the development of the cassette tape recorder manufacture industry in China. Then it moves on to the technology import projects of the sample enterprises during the reform period. Based on the survey information, the chapter gives a general evaluation of the performance of the sample firms in transferring foreign technology.

Chapter 5 analyses the impact of macroeconomic factors on the technological behaviour of the sample state enterprises based on the survey data. It argues that the technological behaviour of the sample state enterprises can be described as a propensity to use capital goods combined with a lack of learning efforts to assimilate the imported technology. There then follows an analysis of the institutional reasons for this technological behaviour. It is argued that the technological behaviour of the sample firms was ultimately defined by changing institutional arrangements governing the operation of SOEs during the different stages of reforms. Furthermore, the technological behaviour of firms was affected by the government's policy on technology imports and by increased product market competition. The conclusion reached is that the ambiguity of the property rights structure led to a deterioration of technological capability of firms.

Chapters 6 and 7 investigate the technological behaviour of Chinese SOEs with detailed case studies of four firms. Chapter 6 reports the process of technology transfer of these firms and the circumstantial context in which they imported and used foreign technology. Among these four firms, two are producers of tape recorders and two are manufacturers of tape decks.[5]

Chapter 7 starts with a general evaluation of the performance of the four sample firms. Then, it presents the technological behaviour of two non-state firms, which are used as a comparison with the state enterprises to show how firms under different institutional arrangements operated differently. The following sections analyse the technological behaviour of the sample SOEs of the case studies against the specific institutional context and market conditions during the reform period. The chapter also shows how government ownership and regulations prevented SOEs from improving their utilization of imported technology. Findings of the case studies supplement the conclusion of the general survey in Chapters 4 and 5 that ambiguity of property right arrangements was responsible for the minimal efforts of the SOEs in undertaking the assimilation of imported technology and led to a deterioration of their technological capabilities.

Finally, Chapter 8 pulls together the major findings of the empirical investigation, points out the limitation of the findings and gives the theoretical and policy implications of this study.

2 Theoretical foundations and research methodology

TECHNOLOGY AND INTERNATIONAL TECHNOLOGY TRANSFER

The nature of technology and technological change

In economics, technology is generally defined as human knowledge applied in production. In a narrow sense, technology is the set of production methods, normally called techniques, available to a production unit. Technology also includes organizational, commercial and managerial knowledge which reflects different ways of organizing production and distribution (Rosegger 1980: 3; Gomulka 1990: 4–6; and Enos 1991: 167). Technology normally consists of many components, embodied in capital goods, in written materials, in people, in firms and in behavioural patterns (Ernst and O'Connor 1989: 20).[1]

Recent research by, among others, Freeman (1974), Rosenberg (1976), Nelson and Winter (1982), Dosi (1984), Pavitt (1985), Ernst and O'Connor (1989), and Nelson (1990a) on the nature of technology has further deepened our understanding of technology and technological change. Technology as the solution to practical problems of production consists of: (1) specific knowledge which is often tacit and generated within the innovative organizations as a result of learning by doing, and (2) a generic understanding about how things work generally.

Many authors (for example, Pavitt 1985: 5; Mowery and Rosenberg 1989: 142–3; and Dosi 1988: 224) argue that technology as the solution to production problems means the rapid adoption of specific knowledge which is mainly learned and accumulated by individual teams through practice. Although many constituent elements of technology are based on widely applicable scientific principles, and technological information is often in written form and even open to the public (at the expense, perhaps, of certain search costs to locate this kind of information), the differing technological capabilities and differing production problems of different firms determine that 'technology is mainly differentiated knowledge about specific applications' (Pavitt 1985: 5). Therefore, an important part of technology is differentiated, tacit, ever-changing and often firm-specific knowledge, accumulated in the process of design, investment, production and marketing.

Although technology is often firm-specific, it may also have the quality of public goods (Gomulka 1990: 25). Firms can draw benefits from the general stock of technological knowledge. They can also learn from other firms through various channels, such as reverse engineering or technology licensing. Indeed, technology cannot be conceptualized as information that is well known and can be costlessly used by all firms (Arrow 1962; Pavitt 1985). But, technological knowledge in modern societies is well articulated and may be codified down to fine detail in the form of blueprints, prototypes, formulas, instructions and designs which can be easily understood by experts. Moreover, even uncodified technology such as skills and know-how can often be learned through personal exchange, learning by doing and watching, and other means. However, the recipient has to possess appropriate technological capabilities and make very strong efforts to master the borrowed technology (Pavitt 1985, Nelson 1990a, and Enos 1991).

Effective technology transfer

The international transfer of technology has played a very important role in the economic development of the world. In the past thirty years, the economic success of the newly industrialized economies (NIEs) has again proved the importance of technology transfer by means of openness and integration with the global economy (World Bank 1991: 88). International technology transfer can be defined as a process in which one kind of technology developed and utilized in one country is transmitted to another country. From the angle of the recipient country, it is reasonable to assume that the ultimate objective in encouraging technology imports is to promote sustained and balanced economic growth (Stewart 1990: 302). To achieve this final goal of economic development, developing countries must acquire and master modern technologies which are vital to productivity increase and industrialization (Ernst and O'Connor 1989: 34). Therefore, the development of technological capabilities is the direct aim of developing countries in transferring advanced technology from abroad.

Technological capability is the ability to utilize technological knowledge effectively. The technological capability of a nation or a firm can be classified into several different areas, each of them representing different levels of technological development. Westphal *et al.* (1985: 171–2) have divided technological capabilities into three broad categories: production capability, investment capability and innovation capability. UNCTC (1987: 34) summarizes several authors' opinions and points out that successful technology transfer should result in the development of technological capability in the acquisition, assimilation, operation, modification, improvement and replication of foreign technology with domestic resources and the subsequent independent development of new technology. Therefore, the process of technological development by means of technology transfer has a cumulative nature and higher levels of technological capability are built on a previously achieved technological base.

Technology transfer as a learning process

The complex and evolutionary nature of technology has important implications for international technology transfer. Many authors argue that international technology transfer is not as easy as the mere purchase of capital goods or the licensing of proprietary technologies. The contribution of imported technology to local technological capability development mainly depends on the efforts of the recipients (Rosenberg and Frischtak 1985: viii; and Nelson 1990a: 72). In the relevant literature, technology transfer has been described as a learning process. Nelson (1990a) uses learning tennis as an analogy to show the importance of active efforts by the learner in the technology transfer process. Although help from foreign firms can facilitate the learning process, it is less important than the active efforts and resource investments of the recipient firms.

Technological capability as the outcome of the technology transfer process can only be obtained by local firms if there is a strong demand for it (Ranis 1990: 161–6; World Bank 1991). The experiences of many developing countries show that government protection for industrial sectors based on the so-called import substitute strategy often leads to unbalanced economic development, inefficient industrial sectors and reliance on foreign technology suppliers (World Bank 1991). As a result of government protection and subsidies, local firms may become rent-seekers rather than initiators of technological change. Therefore, the significant resource investment necessary for the establishment, development and updating of local technological capabilities through technology transfer may be more effective if it is directed by the demand for these capabilities by local firms facing market competition.

Thus, the effectiveness of learning depends to a high degree on receiving correct signals from a competitive market. A market economy is important, because the mastery of modern complex technologies involves implicit understanding, skills, know-how and experience in widely different areas which can hardly be given to and handled by a central planning authority (Hayek 1948/1984: 212; Nelson 1990a: 72). Therefore, properly motivated individual firms coordinated through the market mechanism become the major actors in the process of efficient technology transfer. In most practical circumstances, this is the only way to generate the necessary increments of skill, know-how and experiences which are essential to establish technological capabilities and derive economic growth from imported technology.

THE FORMULATION OF AN ANALYTICAL FRAMEWORK

The technological transfer process and the technological behaviour of recipient firms are influenced by various factors at both macro and micro levels. In order to figure out how these factors affect the technology transfer process of

SOEs during the reform period, an analytical framework has to be established to guide the empirical investigation. This is because a theoretical framework can facilitate the design of research methods, the process of information collection, and the empirical analysis after the relevant data have been gathered (Sanders and Pinhey 1983: 19–20).

Factors affecting technology transfer

In the literature on international technology transfer, a number of factors have been identified that influence the effectiveness of technology transfer by the recipient firms. Many authors mainly stress the factors affecting the costs of acquiring foreign technologies, such as the initial technological capabilities of recipient firms, the role of the governments in negotiating with foreign technology suppliers, and the channels of obtaining foreign technologies, either through foreign direct investment (FDI) or arm's length transactions (Reddy and Zhao 1990; Cusumano and Elenkov 1994). Some authors emphasize the importance of local efforts in learning about foreign technologies (Mytelka 1985). Finally, some authors have identified various factors that influence the learning process of recipient firms. These factors are classified into several categories. For example, Katz (1985) identified four sets of factors: technological absorptive capacity of the recipient firms, competition in a specific market, macroeconomic stability and the nature of the transferred technology. Tornatzky and Fleischer *et al.* (1990) group the factors influencing a firm's path of innovation into three categories: the organizational context, the external environment context and the technology context.

Although the impact of environmental factors on recipient firms is recognized, these authors seldom elaborate on these factors. Furthermore, few economists have pointed out how institutions affect the technological activities of firms in general or discussed the impact of various 'distorted' institutional arrangements on the technological behaviour of recipient firms in utilizing imported technology in particular.

In this study, the focus is on the factors affecting the decision of the recipient firms to make efforts in learning an acquired foreign technology, because the effectiveness of international technology transfer to developing countries depends on the learning efforts of the recipients. Based on the above-mentioned theories, it is argued that the factors affecting the rate and direction of learning efforts of the recipient firms can be divided into four headings: macroeconomic factors; macro-technical factors; microeconomic factors; and micro-technical factors.

Macroeconomic factors mainly influence the general behaviour of enterprises. Factors in this group shape the overall objectives of firms and set the constraints on firms in making business and technological decisions. Specifically, the major factors in this group include the institutional arrangements administering the operation of state firms, the government policy towards technology imports, and the degree of market liberalization. The institutional arrangements administering the operation of state firms define the relationship between the government and state enterprises. They affect the objectives and the decision autonomy of state firms in their business in general and in their technological activities specifically. Technology import policy influences the costs and benefits to state firms of undertaking technological activities. The degree of market liberalization determines the extent of market competition faced by firms. Generally, the more liberalized a market is, the more pressure is put on individual firms to search for alternative technologies (James 1989: 5–6).

Macro-technical factors include technological infrastructure, sciences and technology system, education, and government science and technology policies in general. They mainly influence the appropriateness of the imported technology and the availability of technical services from domestic sources, which in turn affects the choice of appropriate channels of technology transfer and the choice of appropriate technologies (Enos 1989: 9–10; Cimoli and Dosi 1990: 51–71) and hence the performance of technology transfer of the recipient countries (Enos 1989: 20; Westphal *et al.* 1985).

Microeconomic factors mainly involve the competitive climate in the particular market the recipient firm serves, and the management capability of the firm in competing in this particular market. The degree of competition in the particular market may affect the overall performance of firms. Furthermore, in a monopolistic environment, firms are more likely to make efforts to stretch their production output. In a more competitive market, they may be forced to invest in new product development or cost-cutting innovations (Katz 1985: 147–8).

Micro-technical factors comprise mainly the characteristics of the imported technology and the technological capability of the recipient firms before the transfer. These factors obviously influence the learning ability of the recipient firms. Normally, the higher the technological capability of a firm, the more likely it is that the recipient can effectively learn the acquired foreign technology (Cohen and Levinthal 1990; Enos 1991). Thus, these factors determine whether the recipient firm is able to exploit effectively the technological opportunities associated with its efforts to learn the imported technology and to augment its technological capability based on the learning.

Although the factors that influence the learning efforts of the recipient firms are so categorized, this study is not going to decompose the impact of each factor. On the contrary, in this study it is argued that the way through which these factors affect the decisions of the recipient firms to make learning efforts is a complex process in which these factors interact with each other. Furthermore, different factors have different ways of affecting the learning process of technology transfer in the recipient firms, and the magnitude of the impacts of different factors are different, depending on the level and the nature of these factors. Micro factors are normally specific to individual firms and the way through which micro factors impact the learning process of the recipient firms is more direct because they directly affect the ability of the recipient firms to learn the imported technology, the performance of technological activities, and thus the commitment of the firms in making learning efforts.

However, the impact of microeconomic and micro-technological factors on the rate and direction of the learning efforts of the recipient firms is subject to the state of the macro factors that pertain. Among these macro factors, the institutional arrangement governing the operation of firms is especially influential. This is because the behaviour of firms, that is how firms respond to these micro factors, is shaped by the macro environment. If the macro environment remains constant or similar for every firm, the difference in the impact of these micro factors on different firms will probably be significant. If the macro environment is 'distorted' in the sense that the basic institutional arrangements of an economy deviate from that in modern market economies, or if the institutional arrangements of an economy change dramatically, then the impact of macro-socio-economic factors become decisive (see Cusumano and Elenkov 1994).

The process of technology transfer in a changing institutional environment

As discussed above, technology transfer is a complex process involving diverse activities. It is not as simple as the mere purchase of capital goods or the licensing of proprietary technologies (Rosenberg and Frischtak 1985: viii). The effectiveness of this process depends on active learning efforts by the recipient firms (Nelson 1990a). Furthermore, the decision of firms to commit their efforts in learning the imported technology is ultimately affected by the institutional setting of the prevailing property rights structure and government policies. Therefore, in order to investigate the impact of the changes in institutional arrangements as a result of economic reforms on the technological behaviour of state firms, the whole process of technology transfer should be evaluated along with the process of economic reforms.

The analytical framework of the research thus focuses on the changing technological behaviour of state firms as the institutional setting changes. As the basic unit of analysis, the technological behaviour of firms is represented by the efforts the recipient firms make in transferring foreign technology. In turn, learning efforts of the recipient firms are reflected in the technological activities they conduct to learn about the imported technology. Furthermore, the degree to which the recipient firms can independently and efficiently carry out certain technological activities can be used as indicators reflecting the effectiveness of technology transfer (Mytelka 1985: 81). Therefore, evaluating these activities and their performance in different institutional settings will allow us to find out the effects of institutional changes on the technological behaviour of recipient firms. Consequently, the analytical framework comprises the whole technology transfer process of various technological activities proceeding in a changing institutional environment.

The process of technology transfer

Technology transfer is a complex and cumulative process (Pavitt 1985). From the viewpoint of the recipients, this process involves diverse technological activities. Each of them aims at different tasks, draws on different resources and represents different degrees of technological mastery by the recipients in order to achieve the objectives of the recipients at different stages of the process. In this sense, therefore, to evaluate the learning efforts of the recipient firms, it is useful to conceptualize technology transfer as a sequential process of acquiring, implementing, utilizing and improving imported technology.

Many authors have attempted to illustrate the process of technology transfer to developing countries by a dynamic schematic model in which several successive, interrelated and overlapping phases are identified. Enos and Park (1988) identify seven stages in the technology transfer process, according to the function of the activities that the recipient firms carry out at each stage. Lall (1987) describes technology transfer to the recipient firms as a five stage process: project preparation, project execution, product and/or process engineering, industrial engineering, and technology transfer to other firms. His delineation of the stages is based on engineering functions, and he stresses the technological capability gained by the recipient in the process. Dahlman *et al.* (1987) and Westphal *et al.* (1985) divide the process of technology transfer into such economically meaningful categories as pre-investment, investment, production and innovation. The focus of their analysis, however, is on technology acquisition as an investment activity.

The common feature of these models is that they all delineate the process of technology transfer to developing countries as a cumulative sequence, in which the outcome of the previous stage is the basis on which the following stage is built. They all have identified similar stages of acquiring, using and improving foreign technology. However, these models have not stressed the importance of learning in the process of technology transfer. Nor have they explicitly spelled out the difference between the two levels of learning: learning how to use and learning how to improve. As a result, these models overlook an important aspect of effective technology transfer.

In this study, the process of technology transfer is analysed from the standpoint of how the recipient firms learn about the imported technology. That is, how the recipient firms assimilate the acquired foreign technology. Therefore, the process of technology transfer in this study is divided into three broad phases: foreign technology acquisition, foreign technology assimilation and innovation based on assimilated foreign technology. This classification is oriented to local learning efforts in each phase, especially to those in the assimilation phase.

The first phase is technology acquisition. In this phase, local efforts concentrate on selecting and obtaining technology from abroad. The second phase is assimilation of foreign technology. This phase is further divided into two sub-phases. At the 'lower level' of assimilation, the learning efforts mainly focus on how to use the imported technology, such as how to install and start-up the acquired technology, and how to adapt it to local conditions and maintain it regularly to reach normal production conditions. At the 'higher level' of assimilation, the tasks include learning how to modify the imported technology, how to use locally available resources (localization), how to replicate the imported capital goods (imitation) and how to develop indigenous technology based on the acquired technological capabilities.

It must be stressed that the tasks at the 'lower level' of assimilation are the follow-up activities of technology acquisition. They implement the acquisition decision. Although having learned how to use the imported technology provides some necessary experiences for the recipient firms to go further, the stage of learning how to use foreign technology may not necessarily lead to the stage of modifying and improving the imported technology. There can be set-backs, stops and reversals. Therefore, whether the recipient firms can achieve a higher level of assimilation depends on the efforts of recipient firms in actively and consciously learning 'why' the foreign technology works and 'how' to improve it (Nelson 1990a, and Mytelka 1985: 78).

After having mastered the imported technology (characterized as 'major technological changes' in the terms of Enos and Park (1988)), the process of technology transfer will probably enter into the stage of innovation, i.e. the *development* of indigenous technology. Entering this stage implies that the performance of previous stages of technology has been successful. It is the 'ideal' outcome of the process. However, the focus of this study is on the assimilation stage, because the technological capability of the recipient firms to develop new technologies is the result of active learning in the previous stages of technology transfer. Only after achieving the basis of enhanced technological capabilities as the outcome of having assimilated imported technology, will recipient firms be able to undertake indigenous technological innovation.

The process of technology transfer in a changing institutional environment

The purpose of this study is to look into the impact of institutional changes on the process of technology transfer, based on the recognition that the macro-socio-economic environment in China has changed greatly as a result of economic reforms which have inevitably and significantly affected the technological behaviour of Chinese state firms in transferring foreign technology. Therefore, the changing institutional setting of property rights and government policies under which state enterprises operate is considered to be of primary importance. The impact of other factors on the technological behaviour of the recipient firms has been conditioned by the changes in institutional arrangements and can only be appropriately sorted out together with the evaluation of the impact of the changing macro environment.

Therefore, in the analytical framework, the process of technology transfer is integrated into the institutional setting which has been changing over time as a result of economic reforms as shown by Figure 2.1. In this diagram, technology transfer by the recipient firms is viewed as an evolutionary process involving several stages of technological capability establishment and improvement in an evolutionary institutional environment characterized by several major stages of changes.

This analytical framework is, of course, highly abstract. The sequential scheme of technology transfer only shows the 'ideal' outcome of the process. In reality, it is a far more complicated and iterative process (Tornatzky and Fleischer 1990). Although the scheme shows the improvement of the technological capabilities of recipient firms as the result of technology transfer, it cannot be used mechanically as a ruler to measure the effectiveness of specific technology transfer projects. Furthermore, the institutional environment affects the process of technology transfer in a very complex manner. It not only directly impacts the incentive structure and the constraints faced by firms, but also indirectly affects the way in which other factors influence the process of technology transfer.

In this analytical framework, the time dimension is highlighted because of the evolutionary nature of learning involved in technology transfer. In the process of technology transfer, the recipient firms accumulate their technological capabilities through learning by doing technological change in which what they have done in the past affects the alternatives they can choose in the future. Furthermore, this framework stresses not only the impact of institutional arrangements on the process of technology transfer but also significant changes in the institutional arrangements themselves.

The empirical study in this investigation of the technology imports by Chinese state enterprises in the tape recorder industry covers mainly the period from 1982 to 1993. As we will see, in most of the sample firms, the

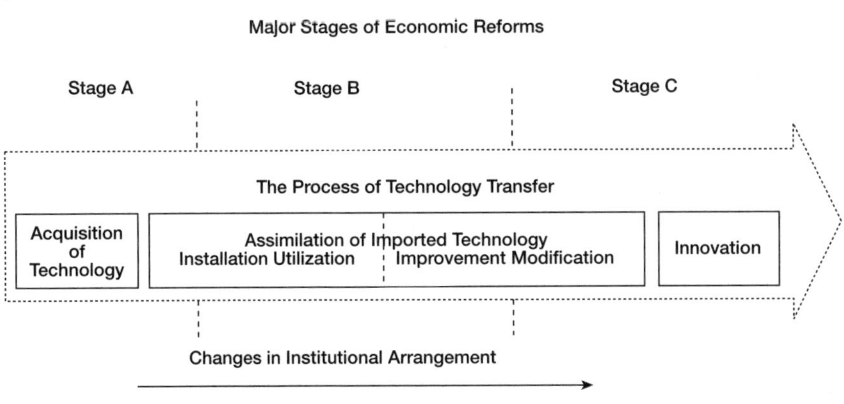

Figure 2.1 The analytical framework

major inputs of foreign technology for the production of cassette tape recorders had been imported and put into production during the mid-1980s under the programme of renovating existing enterprises. The major tasks Chinese state enterprises faced during the second half of the 1980s were how to use and improve the imported technology effectively. In other words, the stages of technology acquisition, installation and utilization of most of the sample firms were in the mid-1980s. The stages of foreign technology utilization and improvement for the sample firms were mainly in the late 1980s and early 1990s.

Therefore, when applying the analytical framework to the empirical study, the early stages of the sample firms' technology transfer coincided with the first stage of institutional change resulting from economic reforms. This stage was characterized by enlargement of the autonomy of individual enterprises in using industrial assets, and by the provision of monetary incentives. Furthermore, as we will see in the following chapters, during this period competition in the market for tape recorders was not strong, because the demand was far greater than the supply.

The assimilation stage of the sample enterprises was mainly in the late 1980s and the early 1990s during which the institutional environment of state enterprises was characterized by the changes brought about by the wide adoption of the Contract Management Responsibility System (CMRS). In addition, as we will see in later chapters, as a result of more firms, mainly the township and village enterprises (TVEs) and foreign direct investment (FDI)[2] enterprises, entering the market for tape recorders, the market became increasingly competitive and the profits of state enterprises were reduced sharply.

The behaviour of state enterprises in transferring foreign technology thus will be analysed in the context of the analytical framework. Specifically, the importation and implementation of tape recorder production technology by state enterprises during the mid-1980s under the technology renovation programme will be investigated against the background of economic reform measures, technology import policy and the market for tape recorders at that time. The emphasis of this study is on the stage of assimilation of imported technology. I will focus on why and how the sample firms made their efforts to learn how to use and improve the imported technology, in the face of changed institutional arrangements and increasing market competition.

THE IMPACT OF INSTITUTIONAL ARRANGEMENTS ON THE TECHNOLOGICAL BEHAVIOUR OF FIRMS

Institutional economics and the property rights approach

The role of institutions in economic life

Institutions as a device of human cooperation have exerted a significant influence on the economic development of society (North and Thomas 1973;

Rosenberg and Birdzell 1986; and Jones 1988). Many economic theorists (for example, North 1981, 1990; Pejovich 1990; and Furubotn and Richter 1991) attempt to integrate institutions into economic analysis. They point out that neoclassical economics fails to deal with current problems of interest to policy makers, because the underlying assumptions, such as zero transaction costs and perfect information, are either overly abstract or unrealistic. In particular, the impact of political, legal, social and economic systems on economic performance has been ignored or regarded as neutral (Furubotn and Richter 1991: 2). The new approach suggested by the so called New Institutional Economics (Langlois 1986; Eggertsson 1990) focuses on how institutional arrangements affect incentives and regulate the economic behaviour of individuals and organizations, as well as on explaining how institutions evolve over time.

Institutions in economics are normally defined as the constraints guiding repeated human interactions in political, economic and social activities (North 1990: 3 and Pejovich 1990: 4). They include both formal rules, such as legal and administrative arrangements, and informal behavioural guidance, such as customs and traditions. The main function of institutions in economic life is to reduce uncertainty in human interaction and cooperation, by providing a framework defining who gets what, who does what and how to carry out interaction and cooperation (North 1990: 3; Pejovich 1990: 1). In a modern market economy, the basic institutions consist of private property rights to production factors, a competitive market mechanism which determines prices and allocates resources, and freedom of contracts (Pejovich 1990; Giersch 1989). All of these are guaranteed by the legal system, and provide incentives to undertake innovation and reduce transaction costs.

Transaction costs

The transaction costs approach is one of the fundamental bases of New Institutional Economics. It postulates that the influence of institutions on economic performance was exerted through their impact on transaction costs and transformation costs (North 1990). Coase (1937) defines transaction costs as the costs of using the price mechanism. In general terms, transaction costs are associated with the costs of acquiring information needed for the execution of exchanges with others. They consist of the costs of measuring the valuable attributes of the thing to be exchanged and the costs of monitoring and enforcing a contract covering the exchange (Williamson 1985: 20–1; Eggertsson 1990: 14; North 1990).

Imperfection and asymmetry of information, and opportunistic behaviour of utility-maximizing individuals, are the fundamental assumptions of modern transaction costs theory (Coase 1937; Simon 1957; and Williamson 1981). According to transaction costs theory, transaction costs are associated with the exchange of property rights from one economic agent to another. Searching for and evaluating the relevance of information take time and resources.

Negotiating and enforcing the exchange of property rights may thus involve substantial costs. Furthermore, utility-maximizing economic agents are always trying to take advantage of information asymmetries. This means that, in the absence of institutional arrangements with both formal rules and informal constraints, free market and voluntary contracts alone cannot guarantee that the exchange of property rights will take place.

Property rights

The theory of property rights is one of the foundations of the New Institutional Economics. Many economists (Cheung 1986; Pejovich 1990; Furubotn and Richter 1991; and Eggertsson 1990) maintain that, if the economic activities of human beings are to be effectively coordinated, property rights must be definable, defendable and divestible. An effective definition of property rights should have the following three elements. First, property rights should be defined exclusively. That means the owner has the exclusive rights to use and to earn income from and to change his property based on his own decisions as long as his action is not against the law and does not encroach on the property rights of others. Second, property rights should be transferable. It is important that an owner has the right to sell or give away his or her property rights voluntarily, because this secures the operation of market mechanisms through which resources can be used by the most productive owners. Third, property rights should be guaranteed by law, which protects owners from being deprived or plundered.

Government regulations

It is generally agreed that governments have played an important role in economic activities through restricting the property rights of owners of the factors of production (Eggertsson 1990: 143–9). Generally, government regulations change the property rights structure to benefit one group of people at the expense of another. Kirzner (1978) analyses the effects of government regulations. He points out that the restrictions of the government on entry of firms into the market and the limitation on profits reduce the possibility that new opportunities will be discovered. Because regulated firms cannot have the same freedom as those firms which are not regulated to cope with uncertainty in the marketplace, it is hard for the regulated firms to compete with other firms on an equal footing. Furthermore, because restrictions on entry or exit lead to a lower degree of competition, they will prevent firms from undertaking entrepreneurial activities (Kirzner 1978).

Principal–agent problem

State ownership often involves a rather tenuous principal–agent relationship, because the management of state owned property is normally conducted through a hierarchical system (Eggertsson 1990: 40–3). A principal–agent relationship is established when the principal delegates some of his or her rights to the agent through a formal or informal contractual arrangement to act on behalf of the principal. In a hierarchical structure, an individual is often a principal and an agent at the same time, as when the property rights are delegated downwards on the organizational ladder.

Because of the asymmetric distribution of information and the opportunistic behaviour of individuals, a contract between the principal and the agent in the real world is far less than perfect and agents have a strong tendency to deviate from the principal's interests at various junctures. This generates 'agency costs'. According to Jensen and Meckling (1976), agency costs are: (1) the monitoring costs to the principal; (2) the bonding costs to the agent; and (3) the residual loss.

In the case of private ownership, although agency costs cannot be eliminated completely, both the principal and the agent have incentives to reduce agency costs through contracts, as long as there is a legal framework to guarantee the implementation of the contract. In the case of public ownership, especially in the Soviet-type centrally planned economy, where ownership is ambiguously delineated and the management of the publicly owned property is not directly constrained by the whole society as the nominal owner, agency costs tend to be large (Leibenstein 1980). There are two major reasons for this. First of all, in the Soviet-type system, the state ownership of enterprises is exercised by government agencies, which delegate parts of the property rights to the industrial assets to the management of state enterprises. In this arrangement, government agencies and enterprises function merely as parts of the administration system of the state (Lee 1991: 26). They are not the residual claimants to the public property of the industrial assets. Therefore, bureaucrats, enterprise managers and workers do not personally bear either the value consequences or the risks associated with their decisions regarding the use of the publicly owned assets (Lee 1991: 26; Wu Jinglian 1993a: 147). In this situation, no one has an incentive to monitor the use of state property.

Second, under this arrangement, the activities of state enterprises are coordinated by a bureaucratic force through a hierarchical structure. In this situation, when the government delegates little decision making power to enterprises, managers and workers have no autonomy and incentive to reduce X-inefficiency[3] and carry out innovations. If the government delegates greater decision power to enterprises through a hierarchical structure, management and workers will tend to expropriate the state owned property of enterprises for private benefit. Because the property rights to the industrial assets of state enterprises are exercised by various governmental organs at various levels, each of these organs acting as the principal has only limited authority and little incentive to monitor the performance of the enterprises as agents. Consequently, the monitoring of the performance of agents is hardly effective.

Common property

Non-exclusivity of property rights is the major feature of common property. In a world where utility maximizing individuals compete with each other for the scarce resources which are not owned exclusively by anyone, this will result in the dissipation of the potential income which could be produced from the common resources (Eggertsson 1990). Furthermore, in reality, optimal use of resources involves adjustments at several margins which depend on the physical attributes of the resources and the institutional arrangements governing the use of the resources. For example, when the resources need time and investment to maintain their productivity, overintensive use and lack of maintenance will destroy the potential value of the resources (Eggertsson 1990: 87–91).

Competition to discover new knowledge

The importance of competition in the learning process is stressed by many economists (for example, Hayek 1978/1984, Demsetz 1982; Kirzner 1992: 139–79, and Katz 1985: 147). Because technology transfer can only be effective through intensive and active learning, market competition will greatly facilitate this process.

The fundamental reason that market competition is generally the most efficient mechanism through which new technology is likely to be generated and learned by participants is that technology is not a kind of static and homogeneous knowledge as assumed by neoclassical economics. Furthermore, in reality, competition does not lead to market equilibrium in which all data are given in advance and the only thing left is to calculate the optimal production quantity. Market competition in real life is a dynamic process in which new knowledge (technology, commercial know-how and human skills to handle technology) is continuously discovered, tested, learned and improved by imitators, and generalized and spread throughout society (Hayek 1978/1984). This process is carried out based on imperfect knowledge and bounded rationality, and the additional data generated by market competition can help to solve information and incentive problems (Hayek 1948/1984, 1978/1984; Klein 1977; Kirzner 1973, 1979).

As far as technology transfer is concerned, competition in domestic and international markets provides incentives for firms to carry out technology transfer. This has proved to be one of the effective approaches for late comers to catch up (Ernst and O'Connor 1989). Moreover, competition is needed to handle the problem of uncertainty arising from the learning process of technology transfer. A central authority can at best only guarantee the fulfilment of whatever is predictable, while market competition forces people to develop and try out their specific knowledge and to rely on the market as the final judge to find which technology is most appropriate to the particular circumstances of time and place in the receiving countries (Hayek 1978/1984: 258).

To keep market competition as a spontaneous order, the institutionally guaranteed freedom of people to enter the market is very important. First of all, free entry can allow individuals to test their own perceptions about new opportunities and make their own decisions to satisfy their ultimate ends. If people are treated discriminatingly by allowing only some groups to enter markets for whatever reasons, knowledge relevant to economic activities cannot be discovered, let alone utilized efficiently. This is because new knowledge and new ways of knowledge are usually perceived by only a few people but would be useful to the whole society if they are made available through competition. That is why Hayek (1973, vol. 3: 49) argues that 'it is the great merit of the market that serves minorities as well as majorities.' Second, free entry of firms (new combinations of production factors) not only facilitates the creation of new values, but also keeps up pressures which force entrepreneurs to innovate further. In the market economy, the monopoly enjoyed by entrepreneurs provides incentives for them to discover new solutions. However, this monopoly should be based only on its economic superiority but nothing else, and must be contestable by other solutions which are more efficient or provide better performance. Contestibility of a monopoly makes entrepreneurs always alert to new opportunities in the market and conveys the information about what their competitors are doing (Klein 1977 and Kirzner 1979). Therefore, keeping the market competitive can 'secure the utilization of widely dispersed knowledge' (Hayek 1973, vol. 2: 117).

The technological behaviour of state owned enterprises

The agency problem and the technological behaviour of state owned enterprises

James (1989: Chapters 2 and 3) explicitly argues that the public ownership is responsible for the technological behaviour of state firms in Tanzania which deviates from the objective of the state technology transfer policy. He maintains that the managers of public enterprises are not 'cosmic maximizers' as assumed by traditional neoclassical economics, who choose technology solely based on the calculation of relative prices (or shadow prices). Rather, managers of state owned enterprises in developing countries have their own objectives, given the institutional arrangements governing the operation of the public sector.

James (1989: Chapter 3) uses X-efficiency theory and the two behavioural concepts of 'engineering-man' and 'bureaucratic-man' to formulate his own analytical framework to explain the choice of technology by state owned enterprises. The central focus of X-efficiency theory is the divergence in the behaviour of a firm's management from cost-minimization caused by the non-coincidence of the objective of the principal (the state represented by the government) and the objective of the agent (the firm's management) (Leibenstein 1980). There are two reasons for this divergence. First, the job contracts between the owner and the managers in firms where the ownership is separated from the management are not fully specified. Managers thus have a certain degree of discretion or autonomy in decision making. Furthermore, in the public sector, the objective of the government is unlikely to be as unambiguous as that of a private owner. This gives more room for managers to achieve their own objectives. Second, there is less competitive pressure on the managers of state enterprises due to protection from the government.

Because of the lack of effort to minimize production costs, the managers of state owned enterprises tend to behave like an 'engineering-man' and a 'bureaucratic-man' as hypothesized by Wells (1975) and Williams (1979) respectively. According to the 'engineering-man' hypothesis, the choice of technology by the managers of state enterprises in developing countries will be capital intensive rather than labour intensive. The reason for the dominance of 'engineering-man' preferences over 'economic-man' preferences is that the competitive pressures faced by state enterprises to minimize costs is weak, because of the monopolistic powers given to state enterprises and because of the ambiguous and inconsistent objectives of various government agencies. As a result, the managers of state enterprises have more opportunities to pursue their own objectives, which mainly involves a preference by managers to use sophisticated modern machinery, because this choice needs less effort in searching for alternative technologies and represents a more prestigious status and less risk.

The behavioural model used by James (1989) tackles the issue of the impacts of institutions on the technological behaviour of firms by relating the choice of foreign technology, one aspect of the process of international technology transfer, to the institutional arrangement of state owned enterprises in developing countries. However, his work mainly deals with the deviation of the efforts of the managers of state firms from cost minimization caused by the lack of the control by the principal and the lack of the market competition. As a result, James does not analyse in depth the impact of property rights on the whole process of technology transfer by state enterprises in developing countries.

Non-exclusive property rights and the technological behaviour of firms

In traditional neoclassical economics, the basic assumption about institutional arrangements is that property rights are perfectly defined and enforced. However, in the real world of modern society, we often find that property rights over some valuable resources are not established for various reasons. Sometimes, it is the high costs involved in defining and enforcing exclusive rights to some natural resources, such as fishery grounds in the high seas, that prevent exclusive property rights from being established. In other cases, technological and economic changes often create new dimensions of property rights or cause uncertainty as to property rights and leave valuable rights in the public domain. Finally, due to ideological or political reasons, resources are sometimes owned by the state. State ownership over production assets dominates in socialist centrally planned economies. However, as mentioned above, state ownership is ambiguous, because property rights to valuable resources are rarely well delineated. Under this institutional arrangement, the property rights to valuable resources are in the public domain to an extent. These resources thus become common property (Eggertsson 1990: 33).

Jensen and Meckling (1979) use the model of property rights-dependent production functions to analyse the behaviour of labour-managed firms, an institutional arrangement prevailing in the former Yugoslavia. They argue that the key to the understanding of labour-managed firms is a clear specification of contracting and property rights in these firms. They point out that the system of labour-managed firms can be described in the following way. The firm is not owned by its employees, although it is called a labour-managed firm. The residual claims on the firm are not held by its employees, in the sense that the rights to use and earn incomes from the resources are contingent on continued employment, and the transfer of these rights through the market is not allowed by law. As a result, the assets of the firm are a form of common property which, together with the other characteristics of modern industrial firms, such as intangibility, indivisibility and depreciability of assets, inevitably leads to the inefficient use of resources.

Cheung (1986) argues that the 'responsibility system' in agricultural reforms of China during the early 1980s actually granted private property rights to farmers through a state lease of land. This is because the rights to use and to obtain incomes are exclusively assigned to farmers. Nevertheless, this kind of responsibility system is unlikely to delineate exclusive property rights to industrial assets because industrial production is greatly different from agricultural production where the property rights to the major productive resource of land are relatively easy to define. Consequently, non-exclusive property rights will lead to inefficiency and the dissipation of value of industrial assets.

Specifically to the technological behaviour of firms, the institutional arrangements of both labour-managed firms and of state enterprises under the 'responsibility system' have the following flaws because property rights to industrial assets of firms are unlikely to be delineated exclusively under these institutions

First of all, property rights over industrial assets are difficult to define through rental contracts or a responsibility system (Jensen and Meckling 1979: 481; and Cheung 1986: 68). This is because a large part of the assets of modern firms are intangible and tacit, and therefore not physically divisible. Moreover, technology embodied in individuals and in the firm as a team is largely firm-specific, and the return on these intangible and tacit assets is difficult to determine in advance. Thus, it is hard to gain access to such assets through rental or 'responsibility' contracts.

Second, a non-exclusive property rights structure will lead to short term behaviour of economic agents. On the one hand, industrial investment projects normally have a long time horizon. On the other hand, the claims of employees on the earnings from the assets of a firm invested with their financial resources is contingent on their continued employment in the firm and they are not allowed by law to sell their claims on earnings in the future. As a result, workers have strong incentives to maximize their near term income by reducing, postponing or eliminating investment in the assets of firm. Alternatively, they will choose projects which can produce high near term incomes (Jensen and Meckling 1979: 482; Cheung 1986: 69).

Finally, since the ability of the firm to generate income cannot be capitalized and traded in the market, the performance of a firm and its management cannot be evaluated accurately and efficiently. According to corporate governance theory, a well organized market in which the performance of a firm is continuously assessed is the most important mechanism for monitoring and controlling the behaviour of managers in modern industrial economies. However, non-transferability of a firm's assets prevents the performance of management from being assessed in this way (Jensen and Meckling 1979: 485).

SOME ISSUES OF RESEARCH METHODOLOGY

A qualitative method

Based on the recognition that the path of institutional change affecting the process of technology transfer is not straightforward and involves iterative interactions among various factors which cannot be controlled in an experimental context, it is not possible to prove rigorously the direction of causality among various factors. Rather, it is intended to provide empirical evidence that supports the argument that institutional arrangements affect the technological behaviour of enterprises in the process of technology transfer. For this purpose, a qualitative method is more suitable than an econometric approach for the following reasons. First of all, although qualitative data are not as precise as quantified data, they are better than quantified data in conveying information of density, vividness and clarity (Patton 1990: Chapter 1; Brewer and Hunter 1989: 160–3). As an empirical study, this research will address the stylized facts reflecting how the technological behaviour of Chinese state firms changes, as the institutional framework has changed as a result of economic reforms. The aim of this study, therefore, is to determine the regularities in the real world which the relevant theory can explain, rather than to pretend to be able to quantify the regularities 'scientifically', but at the cost of employing many unrealistic assumptions (Mayer 1993: 89).

Measurement problems

The analytical framework developed in the previous chapter is the guide to the investigation, because it tells us what to measure and which investigative methods are appropriate (Brewer and Hunter 1989: 21). Based on the analytical framework, the main phenomenon to be measured is the technological behaviour of state firms in using imported technology. The technological behaviour of the recipient firms will be measured from various angles. In other words, a kind of triangulated measurement (Denzin 1978) will be used to evaluate the phenomenon. Based on the analytical framework, the technological behaviour of the recipient firms in using imported technology can be measured along several dimensions, namely (1) the technological activities undertaken by the recipient firms, (2) the objectives of the recipient firms in undertaking these activities, (3) the factors affecting the decision of the recipient firms to commit their efforts to learn the imported technology, and (4) the ultimate performance of these technological activities.

Technological activities, including technological import projects themselves and the follow-up technological activities, reflect the technological behaviour of the recipient firms in transferring foreign technology. In this sense, they are 'behaviors defined in terms of consequences' (Blalock 1979, in Brewer and Hunter 1989: 144–5). This measurement stresses the objective consequences of the behaviour. In this measurement, the function of and the resources invested in a technological activity undertaken by the recipient firm are used to evaluate the technological behaviour of a firm. From the function of a technological activity, the learning effort of the recipient firm can be distinguished from its efforts to achieve other objectives. From the quantity of resources invested in the activity, the size of the effort can be found.

The measurement of the objectives of the recipient firms in undertaking technological activities evaluates the technological behaviour of firms based on their motivations, perception or objectives in carrying out technological activities. From this measurement, we can find the internal states of the firms in behaving in a certain manner. Nevertheless, because of the subjective nature of the measurement, the motivations of firms in technology transfer will be inferred from their goals in undertaking certain technological activities and from their selection criteria for choosing alternatives for a specific technological project. Furthermore, due to the fact that Chinese state enterprises used to operate in a command economy, finding out whether they have decision making autonomy in technological activities will ensure the validity of the measurement, because the more autonomy the state firms have in decision making, the more influential their motivations and objectives are on what kinds of technological activities they would carry out.

The measurement of factors that affect the decision of the recipient firms to commit their efforts to master the imported technology gives an indicator of the linkage between the technological behaviour of the firms and other variables. As we have discussed in this chapter, many factors affect the technological behaviour of the recipient firms. However, the extent to which and the way through which each of these factors influences firms' technological behaviour are different. Because of the nature of the topic, it is difficult to find out the linkage between one particular variable and the technological behaviour of the recipient firms by controlling other variables. To solve this problem, in this study a questionnaire survey has been carried out to find out the opinions of the recipient firms about which factors were important in their decisions to invest in learning imported technology. In addition, detailed case studies are used to investigate which factors were relatively important and in which way these factors affected the technological behaviour of the recipient firms. Although measurements conducted in this way are not very precise in a quantitative sense, they can help us to understand how the real world works.

A performance measurement of technological activities is used in this study to show whether or not a technology transfer process is effective. The purpose of this study is to show how institutional change affects the technological behaviour of individual firms in transferring foreign technology. In other words, it is not primarily statistical but phenomenological. Therefore, the performance measurement is not used to compare the effectiveness of technology transfer in one firm with that in another firm. Rather, the performance measurement is used to show the impact of economic reforms on the process of technology transfer by state firms. Hence, the major concerns of this study are the changes in the performance of technology transfer along the economic reform process in terms of the overall economic outcomes of technology transfer projects, and the extent of technological capability improvement of the recipient firms.

Specifically, three factors are used to evaluate economic outcomes: (1) the time and the resources spent on the fulfilment of concrete technological tasks as compared with the planned budget; (2) the completion of planned targets within deadline; and (3) the production capacity actually reached as compared with the designed capacity. It should be pointed out here that the planned targets are not used as a standard benchmark to judge whether a specific project performs satisfactorily because the planned targets may be unrealistic. Rather, they are used only as a reference to show the changes of overall economic outcomes of individual projects over time.

Improvement in technological capability as a result of technology transfer is indicated by the extent to which the recipient firms actively participate in various stages of the technology transfer process and by the technological activities that the recipient firms can undertake independently (Mytelka 1985: 81; Lall 1987: 17–18). In this respect, successfully carrying out specific technological activities can be used to show an improvement in the technological capability of the recipient firms. These activities may include, for example, the installation and utilization of imported technology in production, the maintenance and normal operation of imported facilities, technological projects to increase the local content of production inputs, modifications made by the recipient firms to adapt the imported technology to local market and raw material conditions, the replication of imported equipment and machinery, quality improvements, and the independent design and development of new products.

Questionnaire interviews and case studies

To conduct qualitative research, a semi-structured questionnaire interview and a detailed case study have been used to collect the necessary information from which the regularities in technological behaviour of state firms can be found. The questionnaire survey has been conducted through face-to-face interviews with managers of state enterprises that undertook technology import projects during the reform era, so that a comprehensive picture of the technological behaviour of various firms can be depicted. Based on an understanding of the general situation, the detailed case studies thus allow the reader to envisage the concrete circumstances under which the firms operated, so that the reader can grasp how and why the firms behaved as they did under different institutional settings.

As discussed earlier, in this study qualitative methods rather than quantitative methods are used, because of the nature of the topic. Therefore, a semi-structured questionnaire interview rather than a statistical survey method has been employed. In other words, this study will not try to find out how significant statistically the linkage between any particular variable and the phenomenon may be. On the contrary, in order to reflect reality, the semi-structured questionnaire stresses the comprehensiveness of the research object, that is the technological behaviour of the recipient firms, by investigating several different dimensions of this phenomenon. Therefore, the information gathered through the semi-structured questionnaire interview will be tabulated and analysed in the context of discussions of overall circumstance faced by the sample firms.

A case study method has been chosen because the research stresses the circumstantial context that exemplifies the institutional environment in which the phenomenon being studied occurred. The disadvantage of the case study method is the inevitably small number of sample firms which may comprise a biased sample of the larger population. Nevertheless, this method can meet the requirements of the research topic, because it gives access to detailed facts that pertain to relatively confined social settings and promises realistic theories that reflect the complexity of actual social life (Brewer and Hunter 1989: 44–5 and 78–9).

In order to collect information for the case studies, the methods of actually visiting the recipient firms and of face-to-face interviewing of managers have been used. These information collection techniques have given the author access to the relevant documents and to interviewees who are familiar with the topic being investigated. Where available, historical documents, such as the feasibility studies and the acceptance reports of technology import projects, and statistics of firms, provide further evidence. In the case studies, the whole process of technology transfer projects is traced. In addition to the questions asked in the questionnaire survey, the face-to-face interviewing of the managers of case study sample firms has included some open-ended questions. These questions focus on (1) the objectives of the firms in their major technological activities related to the imported technology; (2) the institutional settings including the stages of reform and government policies; and (3) the changing conditions of the tape recorder market in which the firms have operated.

The selection of sample firms

The criteria for selecting sample firms include that these firms be representative, comparable and manageable. For the purpose of this study, state enterprises in the tape recorder industry (TRI) located in North and East China have been chosen as the potential sample firms. The field research of China's tape recorder industry was conducted in the first half of 1994. During this field work, forty-two factories were visited. Because some firms either were not willing to cooperate or did not keep adequate information, only thirty-one firms were selected as the questionnaire survey sample firms. Among these thirty-one firms, four were chosen as case study sample firms (see Appendix 1. Sample Enterprises of the Chinese Tape Recorder Industry).

The areas visited included the three municipalities of Beijing, Tianjin and Shanghai, and the two provinces of Jiangsu and Zhejiang, where the tape recorder industry was concentrated, especially in the middle 1980s during which most technology import projects of the industry were carried out. In 1988, these areas together produced 50 per cent of the total output of the country. Another major producer of tape recorders has been Guangdong province, whose output accounted for 30 per cent of the total in 1988. The other twenty-four provinces together produced the remaining 20 per cent. TRI firms in Guangdong have not been visited. This is mainly because most tape recorder firms in Guangdong are assemblers relocated from Hong Kong to exploit the cheaper labour in the mainland of China, while the focus of the study is the changing technological behaviour of Chinese state enterprises during the reform period rather than a comparison between the two technology transfer channels of foreign direct investment and the arm's length transaction by state firms.

The tape recorder industry in China consists of recorder assembly and major components manufacturing. The output of the latter group includes tape decks, magnetic heads and mini electrical motors. The field work concentrated on tape recorder assembly and tape deck production, because magnetic heads and electrical motors are components of tape decks and therefore are not as closely related to tape recorder assembly as tape decks are. Among the thirty-one sample enterprises, there are twenty-three tape recorder assemblers and eight tape deck producers. Normally, tape deck producers assemble tape recorders as well. However, the focus of investigation is on their major business of tape deck production.

The information collected from the field work

The information gathered during the visit to China can be classified into two groups. The first is the background information about government policies and regulations governing technology renovation of existing enterprises with imported technology. The second group of information is the outcome of questionnaire interviews and case study data. It covers the basic information on the sample enterprises, the performance of technology import projects undertaken by these firms, their objectives in undertaking technological activities, their assimilation efforts such as additional technology import projects, training and R&D activities, and the opinions of interviewees about the factors that favoured or discouraged the assimilation of imported technology.

Although the information collected in the field research is abundant, some points have to be made to better understand and utilize it. First of all, information gathered from interviewing managers about the factors affecting the technology transfer process only reflects the present opinions of the interviewees, which may easily be biased, because managers may be tempted to attribute an unsuccessful project to factors beyond their control. Furthermore, managers of state enterprises in China are easily influenced by the official opinions about economic reforms. Therefore, it is necessary to cross-check the information provided by the interviewees against information from other sources. Second, due to the lack of clear definition and precise measurement of qualitative indicators about the technological behaviour of firms, some measurement errors are possible. To overcome this problem, the information about one firm should be compared with that about other firms and the whole industry. Third, the size of interviewed sample firms is not statistically large enough and the control of factors other than institutional environment is not deliberately sought, because the information collected through the field work is not intended to carry out a statistical test to rigorously establish the causality between the institutional setting and the technological behaviour of firms. Instead, this information is used to show how the changing institutional environment influenced the technological behaviour of state firms during the reform period.

3 Economic reforms and technology imports in China during the 1980s

In the 1980s, the Chinese economy experienced a series of reforms that aimed to change it from a centrally planned to a market economy. This process has not yet been completed. Nevertheless, the institutional change has begun to affect the operation of every aspect of the economy. Specifically, technology transfer through importing and utilizing foreign technology by existing state enterprises has been influenced by the process of economic reforms. On the one hand, government technology import policies have changed the role played by enterprises in the acquisition, utilization and assimilation of foreign technology. On the other hand, the change of macroeconomic environment under which enterprises operate has influenced their technological behaviour and affected the overall performance of technology imports.

ECONOMIC REFORMS AND THEIR INFLUENCE ON STATE OWNED INDUSTRIAL ENTERPRISES

The economic system in the pre-reform period

As is widely recognized, the command economic system of pre-reform China was characterized by two features.[1] The first was the omnipotent government planning system and the other was the overwhelming dominance of state ownership of the means of production. Further, the whole economy was organized under the control of a party–government hierarchy. The whole industrial sector was like a single gigantic firm coordinated by bureaucrats through this hierarchical structure. There was no market for production inputs and commodities in which prices were determined by market forces, let alone markets for the production factors of capital and labour. The operation of the economy was carried out by administrative or political forces, rather than under a legal framework. All industrial enterprises, no matter whether they were state owned or collectively owned, were exclusively under the direct administrative control of the government. They were not independent entities with their own autonomy and interests. Rather, they only played the role of a production unit in a large administrative network, and they were merely responsible for transforming inputs into outputs according to the planned production targets given by their supervisory governmental agencies. Production inputs, such as raw materials and intermediate goods, were mainly allocated by the relevant governmental agencies according to the plan, and production outputs were distributed by the government as well. The production factors of capital, labour and land were strictly controlled by the government. Under this system, industrial enterprises performed like cost centres rather than profit or investment centres. They submitted all profits to and got grants from the government.

The central government, through various industrial ministries, owned and directly controlled a proportion of industrial enterprises, while local governments at various levels possessed their own enterprises (Granick 1990). Correspondingly, the national plan, which was highly aggregated, was complemented and disaggregated into local and ministerial plans. The latter two had to be subordinate to the former, but they also embodied the interests of local governments and ministries which actually owned the property. Consequently, the disaggregation of the national plan in pre-reform China was a bargaining and redistribution process between the governments at various levels. The planning of enterprise production was a bargaining process as well. However, the major issues to bargain with were the production targets and raw material allocation rather than profit sharing, because the property rights were owned by the government 'on behalf of the whole people'. The practice of bargaining in planning resulted in the customary way of planning based on previous performance. That is to say that the planned targets and allocations were normally the result of bureaucratic compromises based on the levels achieved in previous years (Byrd 1987: Chapter 5).

The labour system was very rigid in pre-reform China. Only urban residents had the right to work in industrial enterprises.[2] Workers normally did not have the freedom to change their assigned jobs. As a social unit, enterprises performed certain functions of a community or a government including political control as well as welfare provisions. Employees of state enterprises had a lifetime job and comprehensive, though relatively basic, welfare benefits such as housing, health care, education for their children and pensions from their working unit. However, they had little freedom to find alternative employment.

Because capital and labour were strictly controlled by the government, enterprises had no decision making power over wages and investment. The wage rate was determined by the government and was not related to the marginal contribution of labour as a production factor. Enterprises had to hand over all their earnings and the depreciation to the government, which retained the power of allocating capital for new investment projects.

In the pre-reform planning system, enterprises were not the major players in initiating technological change. They had neither the decision making power nor financial resources to undertake new investment projects. Virtually all R&D activities – whether basic research, applied research or even new product development – were controlled and funded by governments at various levels and carried out by scientific and technology institutes separated from industrial enterprises. Scientific and technology institutes carried out research programmes according to the plan and were supposed to transfer their results to production sectors. Under this system, enterprises were not responsible for formal research work and had only a limited role in the later stages of R&D. Most industrial enterprises, except for a few large firms, had no technology development departments. The function of engineers and technicians in industrial enterprises was mainly to guarantee the operation of equipment and the production of existing products according to relatively unchanged technical coefficients.

To sum up, state enterprises in the pre-reform period were simply the agents at the lowest level of the bureaucratic hierarchy and only responsible for transforming inputs to outputs according to state plans. Consequently, the typical behaviour of the state enterprises was to bargain with the planning department for greater resources allocations and lower output quotas, because they had neither the incentives nor autonomy to improve their performance (Byrd 1987: 220–1; Perkins 1988: 603; and Granick 1990). This led to huge residual costs in terms of inefficiency in production and lack of innovation in technology progress. As a result, in the pre-reform period, the growth of China's industrial production was fuelled with heavy capital and labour inputs rather than sustained by increases in total productivity (Perkins 1988: 629; Chen *et al.* 1988; IMF 1991: 2).

Economic reforms in industry

In December 1978, China's leadership announced that a series of economic reforms would be undertaken. The main concerns of the government at that time were to overcome the disastrous outcomes of the Cultural Revolution and to carry out a technologically driven modernization programme. It soon became clear that the fundamental problem in China's economy was the chronically low productivity of labour and capital resulting from the command economic system which made reform of this system necessary.

The fundamental aim of economic reforms since the late 1970s has been to improve the efficiency of the whole economy, especially the efficiency of the state owned enterprises, by 'vitalizing' state enterprises.[3] State enterprise reforms have been focusing on providing more incentives for and delegating more decision making autonomy to state enterprises through separation of their 'management' and 'ownership' functions (IMF 1991: 5; Lee 1991: 33–7; Tam 1992: 9), so that state enterprises could become independent economic entities responsible for their own performance, while retaining state ownership.

However, reform measures were implemented incrementally and unevenly within the original enterprise governance framework designed for mandatory planning through a bureaucratic system (Wu Jinglian 1993a: 147; Jefferson and Rawski 1994). Consequently, state enterprises were unlikely in practice to be responsible for their own performance, especially when they made losses. They did not have the exclusive property rights to the assets, while the government had to retain its control over enterprises. Moreover, the 'management' and 'ownership' of state enterprises could not be separated because state enterprises were still 'owned' by the government (Wu Jinglian 1993a: 168–71; Perkins 1994: 40–1; Wang Dongjing 1995: 35).[4] Nevertheless, as a result of reforms, the relationship between SOEs and the government have changed and the property rights structure of SOEs have been further blurred. These changes inevitably influenced the technological behaviour of enterprises as a collection of managers and workers.

Major content of the economic reforms

Reform measures that affected the behaviour of state enterprises in general and their technological behaviour in particular can be divided into two categories.[5] The first group of reform measures mainly concerns the macroeconomic environment under which state enterprises operate. The second group of reform measures includes those designed to transform state enterprises into independent entities responsible for their own performance. Furthermore, since the early 1980s, the government has encouraged the development of the non-state sector, especially the development of so-called township and village enterprises (TVEs)[6] and foreign direct investment (FDI)[7] firms by giving them preferential treatment. This facilitated the formation of a market mechanism which provided market competition pressures for state enterprises and inevitably affected their behaviour (Jefferson and Rawski 1994).

Reforms of the macroeconomic system included the establishment of market mechanisms for material allocation, reforms of the labour system, and reforms of investment financing. The reform of the price system and the establishment of market mechanisms since 1979 was undertaken gradually through two approaches. The first approach was the administrative adjustment of distorted prices to allow the prices of commodities to reflect their scarcity. The second was gradually to relax the control of prices of certain products, initially consumer goods and then industrial goods, in order to increase the proportion of goods whose prices were determined by market forces. Price liberalization after the mid-1980s resulted in the 'two-tier' price system under which enterprises, when they fulfilled the planned targets, were allowed to sell their above-quota output at higher prices on the free market. As a result of further reform measures, in the early 1990s,[8] the prices of most commodities were determined by the market or only restricted by a price ceiling set by the government.[9]

The reforms of the labour market and the wage system began in 1984 when the total wage bill and bonus fund of enterprises were related to their performance. Within the constraint of the total payroll, management of enterprises was given the authority to decide on the distribution of bonus and wage increases among workers. In 1986, the government introduced a contract labour system which postulated that all new workers would be employed on contracts for a period of three to five years. In addition, new employment opportunities not controlled by the government have been increasing dramatically due to growth in the non-state sectors, such as private and foreign firms. However, since an alternative social welfare system was not established to replace state enterprises in the provision of welfare for workers and the government did not want to cause social instability fed by unemployment, state enterprises were not granted sufficient power to dismiss their employees. Furthermore, the egalitarian distribution of bonus and fringe benefits among workers in the same enterprise still held, although the gap of incomes between different enterprises had become large (World Bank 1990: 163–9; IMF 1991: 7; Harrold and Lall 1993: 41–9; Xie 1993; Perkins 1994: 39).

In order to overcome the inefficiency of investment caused by misallocation of capital and to give more autonomy to enterprises in investment decisions, investment financing of industrial projects had been shifted from the granting of funds from the state budget to bank loans in the mid-1980s. Furthermore, enterprises were required to use a part of their retained profits to finance their own investment projects (Byrd and Tidrick 1987; Wang Jiye 1990). Consequently, the sources of industrial investment financing had changed greatly (World Bank 1990: 96). However, the budget constraints faced by state enterprises had not stiffened, especially in the early stages of the reform. This is because bank loans were still heavily subsidized, especially those for planned investment projects. The allocation of bank loans was controlled by the government at various levels, which had a great interest in promoting investment which could enlarge their revenue base. Investment by enterprises was still subject to pervasive direct or indirect government control. Consequently, enterprises were keen to obtain subsidized bank loans to carry out investment projects, while they tended to avoid financing investment projects out of their own retained funds. If the investment projects turned out to be unsuccessful, enterprises usually could renegotiate with the government for late repayment or concessions. Therefore, the efficiency of investment in state enterprises did not improve very much (Wu Jinglian 1993b).

Enterprise reforms can be divided into three broad periods, according to the major characteristics of changed property rights brought about by the reform measures:

- 1978–83: the pilot experiments in state enterprise reforms, characterized by the focus on provision of material incentives;
- 1984–7: the implementation of state enterprises reform programmes, characterized by the reduction in planning and the sharing of income between enterprises and the government (the replacement of income tax for profits remittance); and

- 1987–the early 1990s: the introduction of ownership reforms, as characterized by the Contract Management Responsibility System.

In the first period, the reforms remained at an experimental level until 1984. During this period, the major idea of enterprise reforms was to give the enterprises more decision making power and to provide enterprises with material incentives which were linked to their performance. The government declared that enterprise management should have autonomy in decisions regarding their daily operations. Meanwhile, the bonus system was re-established through the so-called 'enterprise fund system' which was further developed as an 'enterprise profit retention system' (Lee 1991: 32–3). Under this system, state enterprises were allowed to keep a certain proportion of the profits they generated. The retained profits constituted the 'enterprise fund' which could only be used, according to a certain formula, for production development, employee welfare and the payment of employee bonuses.

However, the profit retention system lasted for only two years. Since the market mechanism had not been established and the budget constraints faced by state enterprises were still loose, the growth of production motivated by retained profits was fuelled by large amounts of inputs, and the product mix of the increased output did not match the needs of the market due to the distorted price system (Wu Jinglian 1993a: 145). Spending on employee bonuses was dramatically increased, leading to significant decreases in state budget revenues (Lee 1991: 34).

The second period started in late 1983 when the state introduced an enterprise income tax system to replace the profit retention system (the programme of tax-for-profits). The idea was to transform state enterprises into independent tax payers and secure revenue for the state, so that the after-tax-profits could reflect the performance of enterprises and the state could have a stable income. This programme consisted of two steps. The first step provided that enterprises hand over 55 per cent of the profits as their income tax. The remaining profits were shared between the government and the enterprises based on various schemes. The second step of the programme was to adopt a normal income tax system with a fixed rate. In addition, an adjustment tax was adopted to compensate for the distorted prices. Under the tax-for-profit system, the retained profits of enterprises decreased greatly compared with the previous years. This is because under the profit retention system, enterprises could obtain a certain percentage of their gross profits, while under the tax-for-profit system, they could only obtain a proportion from their after-tax profit increment. Because of this disincentive, profits generated by enterprises and the government revenue from state enterprises declined dramatically in 1985/1986 (SSB 1991: 212).

Starting in late 1986, in order to reduce the interference of governmental agencies in the daily operations of state enterprises, the Contract Management Responsibility System (CMRS) was introduced. The basic idea of CMRS was that the government as the owner of state enterprises contracted out the right to use the production assets to the enterprises and shared the income from the assets with the enterprises. Under the system, state enterprises signed a contract with their supervisory governmental agencies outlining the responsibilities, obligations, and interests of both parties. By the end of 1987, 80 per cent of state enterprises had adopted the system (Wu Jinglian 1993a: 161), but with variations (for different schemes, see Liu Chao-nan 1993; Chen Derong 1995). The typical scheme widely used by the large and medium-sized state enterprises[10] was the so-called 'two guarantees and one linkage'. Under this scheme, enterprises pledged in the contract to remit a fixed amount of profit with any shortfall to be compensated with the enterprises' own funds. The fixed amount of profits to be remitted was based on previously remitted profits. Once the realized profit exceeded the fixed amount, enterprises would be entitled to have a share of it as specified in the contract. In addition, enterprises had to guarantee the fulfilment of technological advancement targets as set in the contract. Once these two targets were achieved, the total wage bill was linked to the increase in realized profits and taxes, based on the formula that the total wage bill was allowed to increase by 0.7 per cent when the total profits and taxes increased by 1 per cent.

In February 1988, the State Council promulgated a provisional regulation on the CMRS, institutionalizing this system. This clarified the two basic principles of CMRS: (1) the government as the owner contracted out the operation of the assets to enterprises whose full autonomy in management and own interests were guaranteed; (2) enterprises were constrained by the fixed profit remittance as specified in the contract. In addition to CMRS, two sets of responsibility systems were implemented in the mid-1980s to facilitate enterprise reforms. The first was the Manager Responsibility System (MRS) which was meant to separate enterprise management from politics. The second was the Internal Contract Responsibility System (ICRS) which was to break 'the system of eating from a big iron bowl'[11] and to link the income of individual workers to their performance.

Although the government tried various measures to 'refine' the Contract Management Responsibility System (Wu Jinglian 1993a: 164–7), the implementation of CMRS in state enterprises did not improve their performance. Many indicators, such as the increased losses of the state sector (SSB 1991: 410), the enlarged number of state enterprises which suffered losses (Lu Dong 1991), the piling up of inventory of finished goods and the rapid increase of the total liabilities of state enterprises (RMRB (*People's Daily*): 18 and 19 August 1991), showed that the performance of state enterprises deteriorated after the initial spurt of development stimulated by decentralization and market reforms. Entering the 1990s, especially since 1992, a series of experimental reform measures was taken to 'transform the operational mechanism of state enterprises'[12] in order to solve the persistent problem of 'enjoying the benefits but not being responsible for losses' (Wu Jinglian 1993a: 162) in state enterprises. Among these measures, the experiment of transforming state enterprises into joint stock companies is a major one.

In the late 1970s, the government began to encourage the development of industrial enterprises in the rural areas and allowed them to sell their products on the free market. Consequently, individually or collectively owned firms in rural areas grew dramatically. In 1990, township and village enterprises (TVEs) produced 27.4 per cent of the gross value of industrial output (GVIO) in China (SSB 1991: 391). These township and villages enterprises provided a huge number of jobs for the surplus labour that emerged as a result of the implementation of the household responsibility system (see Riskin 1987: Chapter 12) in agricultural production. Because of their private or collective ownership, TVEs normally have a tighter budget constraint than state owned firms. Furthermore, they rely on the free market to obtain their inputs and sell their products. Consequently, they have proved to be more responsible for their performance than state enterprises. Since they are not owned by the government, they enjoy greater autonomy in decision making and suffer less intervention by governmental officials. In addition, they recruit their workers from rural residents for whom TVEs have no obligation to provide lifetime jobs or such welfare benefits as the housing and various subsidies that most urban workers have. Therefore, TVEs enjoy lower labour costs and are more competitive in the market for labour intensive products, while their technological level is normally lower than that of state enterprises, especially in the early years of their development.

The attitude of China's leadership towards foreign direct investment (FDI) changed greatly in 1978, with a view to introducing foreign capital, advanced technology, better management techniques and access to world markets, all of which were badly needed to modernize China's economy. Since then, China has adopted various measures to attract foreign investment. These measures include passage of foreign investment laws, establishment of special economic zones (SEZs), opening up coastal cities to foreign investors, improvement of domestic infrastructure and provision of preferential treatment towards firms with foreign investment.[13] As a result, foreign investment in China has increased steadily. From 1979 to 1990, China approved foreign investment in more than 29,000 firms and attracted foreign capital in excess of US $40 billion (SSB 1991: 629). As compared with domestic firms, FDI firms had many preferential benefits, such as low tax rates and long tax holidays, the right to engage in foreign trade, exemption from tariffs for imported capital goods and for imported inputs, if the latter were used for production of exported goods, etc. On the other hand, FDI firms normally did not enjoy easy access to subsidized inputs and credit and were required to balance their own foreign exchange earnings and expenditures (UNCTC 1988). Furthermore, unlike state firms, FDI firms have no government agencies as their bosses and, therefore, are less subject to official interference. They hire their workers from the labour market, which has become more liberalized in recent years. FDI firms do not guarantee lifetime jobs, nor do they provide such welfare benefits as housing and various subsidies. As a result, FDI firms have behaved more responsively to market signals.

The development of TVEs and FDI firms represented a unique characteristic in the transformation of a centrally planned economy into a market economy in the sense that these non-state owned entities both contributed to the economic growth during the transitional period and facilitated the formation of a market mechanism. The competition of these firms for production factors and for market share inevitably affected the performance of state enterprises. The fast growth of the non-state sectors of TVEs and FDI firms eroded the dominance of state enterprises in total industrial production.[14] Consequently, the monopolistic position of state owned enterprises was undermined and their profitability was reduced (Jefferson and Rawski 1994: 58–62). Competition from the non-state sector forced state enterprises to pay more attention to market signals (Jefferson and Rawski 1994: 60). However, due to the continued absence of well-delineated property rights arrangements, competition from aggressive non-state sector firms did not motivate state enterprises to improve their efficiency.[15]

Influences of economic reforms on state industrial enterprises

Economic reforms and state enterprises' performance

Evaluating the impact of economic reforms on the performance of the state sector of industry is a difficult and sometimes controversial task. However, the overwhelming opinion, including that of Chinese official commentators, is that the performance of state firms is still not satisfactory in terms of the goals of the government, despite fifteen years of reforms (Wu Jinglian 1993b; Tam 1992; World Bank 1990; Perkins 1994). First of all, state enterprises are far less dynamic in development than the non-state sector has been. Although the growth of real industrial production during the reform period has been faster than over the period from 1966 to 1978, this is mainly attributable to the rapid growth of the non-state sector, such as collective, private and foreign firms. (SSB 1991: 395). Second, although some authors (Chen *et al.* 1988; Dollar 1990) find that the growth rate of total factor productivity in Chinese industry, including the state sector, at the early stage of reform was higher than that before 1978, other commentators (for example, Jefferson *et al.* 1992) show that the growth rate of the state sector's total productivity is low compared to that of the non-state sector. Perkins (1988) argues that the large and medium-sized industrial enterprises may be the only ones that have not experienced total productivity gains during this period. That means that the growth pattern of state enterprises is still heavily reliant on the increase in production factor inputs rather than on technological progress in production and improved management.

The deterioration of the performance of state enterprises can be explained superficially in terms of the difficulties caused by the macroeconomic environment and the lack of autonomy of enterprises. Among many authors, Zhou Shulian (1987), Lu Dong (1991) and Wu Shuqing (1991) correctly analyse the major difficulties and problems met by state enterprises in recent years. Although the autonomy rights of enterprises in operation have been formulated in the Law of Enterprises, they have not been completely put into effect. The heavy burden of taxes and various levies, together with a low official depreciation rate, prevents enterprises from undertaking replacement and investment projects to renovate their equipment. Numerous administrative interventions and inspections from various government organs not only impede managers in making their own decisions, but also reduce their attention to managerial issues (World Bank, 1990: 174). Despite the various reform measures regarding the labour system, state enterprises still have not been granted the right to streamline their over-staffed labour force. Moreover, they continue to have an obligation to provide social welfare for their lifetime employees and families. The function of providing housing, health care, and fringe benefits absorbs a large amount of resources and prevents improvements in productive efficiency. In spite of these difficulties faced by state enterprises, many authors argue that the short term behaviour caused by the partially reformed economic system is one important reason for the unsatisfactory performance of state enterprises in recent years (Xu and Chen 1991; Lee 1991; Tam 1992; Wu Jinglian 1993a; Harrold and Lall 1993).

Economic reforms and enterprise behaviour

The state enterprise reform during the decade from the early 1980s to the 1990s was mainly implemented through enlarging the decision making autonomy and providing financial incentives within the framework of a command economy (Wu Jinglian, 1993a; Jefferson and Rawski 1994). As a result, new institutional arrangements came out from the hierarchical system of SOE governance. Therefore, Chinese SOEs had several basic characteristics which did not change very much as a result of the reforms. First of all, they still belonged to or were supervised by at least one governmental agency which was either at the central level (industrial ministries) or at the local level (provincial or municipal industrial bureaux), while they were also subject to intervention by other government agencies. Second, the production factors of land, labour and capital were allocated through the hierarchy of the government administration system. Once these resources were allocated to enterprises, they were very difficult to re-allocate to other uses. Third, the majority of industrial enterprises, especially those small and medium-sized ones, comprised of only one single factory. Due to the lack of coordination of production among firms located in different regions or belonging to different governmental agencies through the planning system, the enterprises were relatively unspecialized and tended to carry out as many production functions by themselves as possible (World Bank 1990: 171–2). Finally, SOEs also assumed many social responsibilities for their lifetime employees ranging from housing, canteens, medical care, pensions (which was paid out of the current revenues of enterprises), and kindergartens and schools for children of their employees (Harrold and Lall 1993). Consequently, a large proportion of the assets of SOEs was unproductive (Jefferson, *et al.* 1992: 244–5).

Although some basic characteristics of SOEs were unchanged, economic reforms after the late 1970s enlarged the decision making autonomy and the macro environment of enterprises. These reform measures inevitably affected the behaviour of state enterprises. However, the change in enterprise behaviour evolved gradually and the degree of change varied greatly across regions and sectors (Jefferson *et al.* 1992: 7). This is because a blueprint for the implementation of a well designed reform programme did not exist. Instead, a pragmatic trial-and-error approach towards the establishment of a market mechanism was adopted (IMF 1991: 3). As far as the macroeconomic system is concerned, those areas where reform measures were relatively easy to implement were reformed first. Similarly, reforms in different industries were not even. The production of consumer goods was liberalized earlier than heavy industry, because the impact of reforms in the former was relatively easy to control. Reforms among different regions were also uneven, with the coastal open areas, especially SEZs in southern China, being in the lead.

Consequently, the reforms created more dynamic response from collectively and privately owned firms, because they were less controlled by the government and they had to rely on themselves for survival. The impact of economic reforms on state enterprises was more complicated than on non-state firms. As we discussed before, enterprise reforms were mainly carried out within the framework designed for mandatory planning through providing more incentives for and delegating more decision making autonomy to enterprises. Consequently, the inherent ambiguity of property rights under state ownership was further blurred. In this situation, it was impossible to separate the functions of 'management' and 'ownership' of state enterprises (Wang Dongjing 1995). This is because as the actual owner of state enterprises, the government had to monitor the operation of enterprises to ensure its assets were being used properly. Therefore, it retained the most important rights, such as the appointment of managing directors of enterprises, the disposal of enterprise assets, foreign trade and important financial matters (Qian Yingyi 1995, Zhu Tao, 1993). This led to continued government intervention into the management of enterprises.

As a result of economic reforms, state enterprises were gradually granted more and more autonomy in their daily operation and asked to assume more responsibility for their performance and their long term development. In addition, the macroeconomic environment under which they operated had changed greatly. As a result, they became more responsive to market signals than before. Therefore, it can be argued that the changing behaviour of state enterprises is a combined result of the enlarged autonomy of enterprises, the emergence of market mechanisms and the legacy of central planning. Furthermore, the behaviour of state enterprises evolved during the reform period with the impact of central planning decreasing and the impact of market forces increasing, as the reforms progressed. Nevertheless, because of the remaining state ownership, state enterprises had not become independent entities responsible for their own performance (Wu Jinglian 1993b). Under the institutional framework in which the property rights to the assets of state enterprises became increasingly ambiguous, managers and workers as the agents of the government tended to obtain a larger share when they made profits and to expect the government to bail them out when they made losses (Wu Jinglian 1993a; Wang Dongjing 1995).

The behaviour of enterprises is ultimately determined by the institutional arrangements under which they operate. Because the economic reforms in China are incremental by nature and unevenly implemented, it can be argued that the resultant change of enterprise behaviour evolved through two broad stages. The first stage, characterized by the trend of worker income maximization, emerged in the early and mid-1980s (Walder 1989). The second stage was characterized by short term behaviour due to the common property problem[16] resulting from implementation of CMRS after 1987 (Wu Jinglian 1993a). The discussion below analyses how economic reform measures at different reform stages affected the behaviour of state enterprises, with the focus on the potential relevance of the changing behaviour to the process of technology transfer in state enterprises.

The emergence of employee income maximization

Before the reforms, the behaviour of state enterprises can be described as follows. Under the system of command central planning, the response of state enterprises to planned targets was basically passive, with little room for them to make their own decisions. Consequently, enterprise behaviour before the reforms was limited to bargaining with their superior governmental agencies for better treatment in terms of a larger supply of inputs and lower planned production quotas. In addition, they often tried to establish informal links with other organizations in order to reduce the uncertainty created by ineffective planning (Byrd 1987: 84–7, 219).

In the period between the early and mid-1980s, enterprise reform was characterized by delegating more decision making autonomy to and providing more monetary incentives for enterprises within the original central planning framework (Wu Jinglian 1993a: 142). Because the reforms were carried out by gradually increasing the unplanned market exchange in order to motivate enterprises to produce more, a 'two-tier' system gradually emerged (Rawski 1992; IMF 1991: 6; and Byrd 1987: 354–90). Under this system, state enterprises could sell their above-plan output at the higher free market prices.

Byrd (1987: 362–5) and Wu Jinglian (1993a: 144–51) have analysed the reasons for this type of enterprise behaviour. They argue that under the central planning system, the demand for resources normally far exceeded supply, which generated a persistent seller's market in which the planned prices were much lower than the market clearing level. The emergence of the 'two-tier' system in which the above plan output was allowed to be sold on the free market actually 'monetized' the difference between the planned prices and the free market prices. Due to the implementation of the profits retention system, state enterprises now had a strong motivation to generate more profits. In this situation, the easiest path for state enterprises was to reduce the share of planned output and expand the production of goods which could easily be sold on the free market, and at the same time to obtain more allocated resources from the planning system. This was doubly beneficial because enterprises which were allocated resources were actually subsidized by the planning system, while enterprises which could sell their output at the higher free market prices actually avoided handing out the 'rents' monetized by the 'two-tier' system.

As is mentioned above, economic reforms were carried out unevenly among different areas and different sectors. Generally, the production and marketing of consumer goods were liberalized earlier and to a larger degree than those of raw materials. Similarly, the market mechanism for products was established earlier than that for production factors. Consequently, in practice, state enterprises and their supervisory local governments, which now were delegated more decision making authority, tended to bargain with their respective higher authority for more bank loans and larger raw materials allocations and to expand the production of those manufactured goods whose prices had been liberalized.

As we will argue in the later chapters, this behaviour is most likely to have influenced the technological behaviour of state enterprises in importing foreign technology. First, technology imports of enterprises have tended to concentrate on the production of certain products which are more lucrative than others. This has caused so-called duplicated imports of similar technology. Second, enterprises have tended to spend money on importation of production facilities to expand their production capacity, rather than on technical know-how or on assimilation of imported technology.

Development of short term behaviour since the late 1980s

As a result of the wide implementation of CMRS in the late 1980s, state enterprises were granted further autonomy in decision making, within the restrictions allowed by the contractual obligation to remit profits. It was hoped by the central government that this arrangement could provide incentives for enterprises to improve their performance (Wu Jinglian 1993a: 161). However, because of the inherent difficulties of CMRS in delineating property rights to industrial assets,[17] it can be argued that state enterprises tended to behave with a short term orientation. That is to say, under this arrangement, state enterprises were reluctant to invest their own retained funds for long term development, although long term development was their responsibility as stipulated in the contract. On the contrary, they tended to maximize the near term income of employees. This was normally done either through expanding output by increasing inputs or through diverting retained profits from investment uses to use them instead to finance bonus or welfare funds. This behaviour inevitably sacrificed the long term development of enterprises. Consequently, although the budget constraint faced by state enterprises under CMRS was tighter than before (Jefferson and Rawski 1994: 52), state enterprises were still not pressured to take responsibility for their long term performance.

The short term behaviour caused by CMRS is well recognized (Wu Jinglian 1993a; IMF 1991: 6; Liu Shijin 1993: 15–18; Xu and Chen 1991, Wu Shuqing 1991). Observers argue that CMRS caused the following problems. First, the system actually strengthened the power of supervisory agencies, because supervisory agencies had not only the authority to grant contracts, but also the right to monitor the implementation of contracts and the power to change their terms. As a result, the intervention of the government in the operation of enterprises was inevitable.

Second, CMRS was a compromise between the reform of property rights and the conservation of existing institutions. It was easier to insist on the performance of contractual obligations when macroeconomic conditions were stable. However, this was inconsistent with the institutional change inevitably caused by reforms. In addition, the widespread adoption of CMRS implied that all enterprises had to survive and remit profits, regardless of whether their products were in shortage or oversupplied. Therefore, the CMRS tended to increase the difficulties in further reforms of the economic system and in changes of the unbalanced industrial structure caused by the distorted price system and government protection.

Third, under the arrangement of CMRS, state enterprises did not become independent economic entities with their own property. Instead, they were still agents of the government. As a result, they were only interested in the fulfilment of contracted targets, mainly the profit remittance target, because it was a measurement of performance closely related to their own benefits (in terms of the income of workers and managers). Because the contracted targets were normally determined arbitrarily through negotiation between the enterprise and its superior(s) due to the distorted price system and other macroeconomic conditions, and because the contract term was normally three to five years, enterprises tended to maximize their retained profits through bargaining with superiors for low realized profit targets and low percentages of remittances to the government. Furthermore, they tended to seek short term outcomes through increasing inputs and outputs and had no incentives for long term investment in technological progress.

Finally, CMRS did not greatly change the system of lifetime employment and enterprise-based welfare programmes (including housing and pensions) in state enterprises. This permanent employment allowed workers to press management for higher incomes which were usually distributed in a more or less egalitarian fashion (World Bank 1990: 163–4). Due to political reasons, local governments actually tacitly consented to payment increases in state enterprises to compensate for increases in the cost of living (IMF 1991: 7). Because the welfare of housing and pensions was provided by enterprises rather than by the government through social security programmes, this rigid labour system restricted labour mobility. Furthermore, this system put unequal burdens on old enterprises which had to pay a large pension bill for their retired workers. It was also one of the most important reasons for the difficult implementation of a bankruptcy system in state enterprises (World Bank 1990: 237; Xie 1993). Under this labour system, the even income distribution of workers provided little incentive for better performance. Finally, investment in production development was often sacrificed to increase the share of bonus and welfare payments in the retained profits (Liu Shijin 1993: 14–16; World Bank 1990: 163).

Walder (1989), Huang (1990) and Lee (1991) extend their analyses of Chinese state enterprises by including the behaviour of the government organs which supervised state enterprises. They argue that collusion among supervisory government agencies, enterprise managers and workers was responsible for the poor performance of state enterprises and that this collusion was created by the semi-reformed economic system. Because the managers had no power to fire workers, in order to gain their cooperation and support, enterprise managers had no choice but to yield to the pressures from below for more bonuses and fringe benefits. On the other hand, state enterprises were the major sources of revenues and the base of powers of the supervisory governmental organs. Keeping their own enterprises growing was consistent with their interests. The usual and easiest way for them to do that was by providing bank loans and giving preferential policies to their enterprises (Huang 1990: 446–53). Due to the cellular structure of administration system,[18] local branches of banks were under the control of local governments. The preferential policies were often at the expense of the central government and other localities (Huang 1990).

Implications of enterprise behaviour change for technological investment in state enterprises

Given the semi-reformed enterprise governance system and macroeconomic environment, the incentives for state enterprises to carry out technological projects were unlikely to have been intensified (Jiang and Shi 1993 and Conroy 1992: 33–7). Therefore, it can be assumed that the technological behaviour of state enterprises under the partly reformed system represented by CMRS was unlikely to promote technological change for the following reasons. First, investment was still controlled directly and indirectly by the government that usually had objectives and priorities different from those of enterprises. Technological change often requires a large amount of investment. Given the lack of alternative financing sources and inadequate profit retention, individual enterprises, especially those whose equipment and products were outdated, probably did not have adequate financial ability to carry out large projects.

Second, although prices of most commodities were increasingly determined by the market forces of demand and supply, state enterprises were most likely to be less responsive than private and collective enterprises to competition in product markets, because state enterprises in general had easier access to government protection and subsidies. They could use price increases to compensate for cost increases, if their market was protected. Even in sectors where market competition was much more developed, market pressure hardly provided an incentive for promoting technological change, because state enterprises could rely on the government to bail them out. Under this situation, the pressure for cost reductions through technological change in state enterprises was weak (Conroy 1992: 35–7).

Third, investments in technological renovation projects and new product development are riskier and more long term oriented than those to expand production of existing products. Investment in technological change projects also consumes a considerable proportion of resources available to the enterprise and sacrifices the current profits in return for long term development. Given the ambiguous property rights relations during the CMRS period between enterprises and their governmental superiors regarding the production assets established through long term technological efforts, state enterprises were unlikely to engage in technology oriented investment projects with their own funds (Wu Jinglian 1993a: 163). On the contrary, enterprises tended to compete for the investment projects funded by the government, regardless of the rate of return on investment of these projects, because the repayment terms were usually flexible and renegotiable (Jiang and Shi 1993).

TECHNOLOGY IMPORTS IN CHINA DURING THE 1980s

Technology imports of China before the reform

The desire for advanced technology from abroad was probably the most important consideration when China switched to her 'open door' policy in 1979. However, this new policy was not the first attempt to acquire modern technology from foreign countries. In fact, imported technology has been playing a decisive role in China's drive to industrialization since the early 1950s.

Nevertheless, the inflow of foreign technology over the period from 1950 to 1980 fluctuated distinctively, reflecting the changes in the domestic political arena and international environment. The history of foreign technology importation during the pre-reform period can be divided into four phases (Barnett 1981; Ho and Huenemann 1984; and Conroy 1986). The first phase of technology transfer from abroad started in the early 1950s when the new communist government adopted a 'lean-to-one-side' foreign policy. The Soviet Union was viewed by the leadership as a model to learn from and a reliable source to get economic and technological assistance. The strategy of the government was to industrialize the economy previously dominated by traditional agriculture as soon as possible through the importation of large turn-key projects and Soviet equipment. As a result, large scale technology transfer was carried out throughout the whole of the 1950s.

During this period, China planned to build or renovate 304 turn-key plants with complete sets of equipment imported from Russia and Eastern European countries. By 1960 when the Soviet aid was halted, a large part of these projects was completed. In addition, China received from the Soviet bloc her most needed technical assistance in the form of technical services of engineers and technicians, training of Chinese technical personnel and provision of technical documents and blueprints for machinery, equipment and plants (Ho and Huenemann 1984).

The massive transfer of technology from the Soviet Union during the 1950s formed the technological core of China's modern industries. Based on it, China established her own industrial facilities within a decade. With the Soviet help, which stressed the importance of heavy industry and machine building, China gained the technological capabilities of designing and manufacturing sophisticated modern machinery, such as machine tools, electricity generators and equipment for heavy industries.

The experience of technology acquisition during this period had a profound influence on technology imports later on and resulted in the conceptual misunderstanding of the Chinese leadership about technology transfer. Since large turn-key projects accounted for the most important part of technology transfer during this period, machinery and equipment were viewed by government officials as equivalent to technology itself. Based on this understanding, the most effective and simplest way to narrow the technological gap between China and industrial countries was to buy advanced equipment and machinery.

The second phase of technology imports represents an attempt by China to search for alternative sources of technology from the West instead of the Soviet bloc (Conroy 1992: 182–92). Negotiations with Japan and Western European countries for seller's credit to import complete sets of equipment began at the end of 1962. Starting with a small amount, the scale of machinery and whole plant imports increased rapidly until 1966 when the imports of foreign technology were interrupted by the outbreak of the Cultural Revolution (1966–76). After the violent peak of the Cultural Revolution, the moderates in the leadership managed to gain influence in policy making and resumed imports of equipment from the West. Therefore, the third phase could be viewed as the continuation of the second phase. In 1973, the leadership approved the proposal of the State Planning Commission to spend US $4.3 billion for importing complete sets of equipment and turn-key plants from abroad over the following four years.

The fourth phase of technology importation started immediately after the announcement by the leadership that the Cultural Revolution had ended and economic development would be the first priority of the government (Barnett

1981; Conroy 1992: 182–92). The new economic development strategy depicted in the ten-year plan was extraordinarily ambitious and concentrated on industry, especially heavy industry, including iron and steel, energy and chemical industries. Once more, the leadership tried to accomplish the targets of the new strategy through the purchase of modern equipment and turn-key plants from the West. Nevertheless, this ambitious modernization programme further strained domestic resources and was bound to fail. As a result, a more pragmatic economic policy was announced in 1979 and technology imports declined dramatically until 1983.

During the three decades before the early 1980s, China had built a sizable industrial foundation. However, this accomplishment was achieved through the mobilization of enormous resources. Since the government intervention in economic development was heavily affected by the political struggles within the leadership, the resources were not efficiently utilized and the many industrialization programmes were not effective as they were planned. Furthermore, the central planning system was rigid and unable to promote technological progress. As a result, the gap in technology between China and industrial countries has not been narrowed. In the late 1970s, it was recognized by the leadership that to improve the productivity of the economy the economic system of central planning had to be reformed.

Government policy and the administration of technology imports in the 1980s[19]

In the 1980s, along with the economic reforms, the Chinese authorities changed the technology import policy to focus more on technological renovation of existing enterprises. To respond to this new policy, the State Economic Commission formulated and implemented the programme of renovating existing enterprises with imported technology within the Sixth Five-Year Plan period. The fundamental reasons for the new policy can be found in the attempt to reduce inefficiency due to the central planning system. However, the immediate trigger for the implementation of the policy was the failure of the over-ambitious economic development plan of 1978, which was characterized by a large number of whole plant import contracts. It came to be recognized that China could not be modernized rapidly through the construction of a number of large scale projects with imported complete sets of equipment and that the overheated purchasing of foreign machinery and equipment had resulted in a serious foreign trade deficit. The government, therefore, decided to carry out a three-year plan from 1979 to 1981 to readjust the national economy (Barnett 1981: 83–5 and Tung 1982: 51–9). As a result, imports of machinery and equipment decreased dramatically until 1982 (MoFERT 1990: 296).

The most important impact of adoption of the new economic development strategy on technology transfer was the reassessment of previous policies on technology imports. The policy makers identified several drawbacks in previous policies: (1) an over-reliance on imports of complete sets of equipment and duplicated imports of whole plant; (2) neglect of software, such as patents, know-how, and technical documents, in technology imports; and (3) a neglect of technological updating of existing enterprises (Conroy 1992: 193).

As a result, in the early 1980s, China changed her industrialization policy from an 'extensive' growth strategy (relying on establishing new plants) to an 'intensive' one (technologically updating existing enterprises).[20] Consequently, the focus of the new technology imports policy shifted from construction of large turn-key projects, which was the normal practice of the 1960s and 1970s, to the importation of key pieces of equipment and product technologies to re-equip existing industrial enterprises and to update their products. It was argued by Chinese policy makers that this approach would facilitate the 'technological progress of enterprises' and improve the effectiveness of technology import projects at lower cost, because the size of such investments is small, the construction period is short and thus the outcome can be achieved quickly (Sheng *et al.* eds 1991).

The new plan of technology imports to renovate existing enterprises emerged in 1982. It covered the last three years of the Sixth Five-Year Plan from 1983 to 1985 (Sheng *et al.* eds 1991: 86–8), and proposed to import 3,000 technological projects at a total cost of US $3 billion. Under this plan, technology imports were to be combined with the technological renovation projects of existing enterprises. In line with the reforms to decentralize decision making powers and to handle the numerous projects effectively, a policy of relaxing central control and enhancing local initiatives was adopted. The planning, approving and implementation of technology import projects was carried out by a multi-level management system coordinated by the State Economic Commission (SEC), in order to simplify planning and approval procedures. The SEC is responsible for formulating the priority sectors and regions, and, together with other central economic agencies, is responsible for the approval of large projects. Local governments were delegated planning and approval authority for projects below a certain amount of investment. In 1983, the delegation of authority regarding technology imports was tried in Shanghai and Tianjin, which were granted the power to approve import projects whose total investment was below the norm set by the central government. Later on, the coastal 'open' cities were granted similar authority (Conroy 1992: 196–200). The funds for projects to renovate existing enterprises with imported technology were allocated to localities, which were granted approval authority along with ministries. In addition, localities and ministries could retain a certain proportion of their foreign currency earnings to finance their own technology import projects (SEC 1985: 44–50). This mechanism of allocating funds has two characteristics. First, the funds were administratively allocated, and although direct grants were replaced by bank loans, these as well as foreign exchange quotas were still subsidized (SEC 1985: 51–5). Second, approval power was decentralized, primarily to governments at various levels, as the funds were allocated through the administrative hierarchy.

In the programme to renovate existing enterprises with imported technology, priority was given to disembodied technology, key pieces of equipment and assembly lines. The argument of central policy makers for this is that disembodied technology, such as patents, know-how and blueprints, cost less and was more effective than simply importing capital goods in transferring technology. The key pieces of equipment could be put into production quickly and investment could be recovered within a short period of time. Importing assembly lines and complete knocked down (CKD) kits to assemble finished goods, however, was a way of technology transfer suited to sectors such as the electronics industry, especially consumer electronic goods. It would allow Chinese enterprises to meet the rapidly increasing domestic demand for these products, because local enterprises had no ability to develop their own new products (Sheng *et al.* eds 1992: 88). It was, in fact, an import substitution strategy.

In order to save foreign currency in the importation of CKD kits for imported assembly lines and of duplicated equipment, the central planners launched the 'localization' programme[21] to help domestic firms to establish parts production capability for imported product technology and to stop the duplicated imports of similar capital goods. A typical example of the localization programme is the 'twelve series programme' initiated and coordinated by the State Economic Commission in 1986. Individual provinces and industrial ministries also organized similar localization programmes (Sheng *et al.* eds 1991: 88). The 'twelve series programme' included localization projects for production lines for beer, processing equipment for garments, numerically controlled machine tools, colour televisions, fertilizers, coal mining equipment, etc. (Sheng *et al.* eds 1991: 89). This programme was organized and coordinated through industrial ministries. Each area of technology to be localized was broken down into several projects which was then allocated to enterprises with appropriate technological capability. To promote the localization of imported technology, technological renovation projects within the programme were brought into line with the national plan. In the following years, RMB ¥2.3 billion of subsidized bank loans and US $230 million in foreign exchange were provided in the national plan (Sheng *et al.* eds 1991: 89).

The aims of the technology imports programme in the 1980s were to substitute imports and promote exports to earn foreign currencies which were badly needed to buy foreign equipment and machinery. Therefore, the priority of technology imports was given to the coastal areas and to the textile, light industry, electronics and machinery sectors (Conroy 1992: 196–205). Coastal cities were not only granted independent planning authority, but also larger shares in the whole budget of technology imports.[22] As far as sectoral priority is concerned, a programme called '1200 light industry and textile projects' was implemented to update these two traditional earners of foreign currency. The manufacture of machinery was considered to be the base for indigenous technological development, since it was the sector that provided equipment and components for other sectors. Electronic products, especially consumer electronic goods, were in short supply. A programme called '550 machinery and electronic projects' was launched to provide substitutes for imports (Sheng *et al.* eds 1991).

In line with the enterprise reforms, the autonomy of state enterprises in decision making regarding technology imports was increased. They began to play an active, though limited role in the decision making process, especially in relation to the specific types of technology to be imported (Grow 1987). Nevertheless, throughout the 1980s, they still relied largely on funds supplied by the government, especially for foreign currency. They were also subject to the approval of supervisory agencies, which was often a lengthy and over-elaborate process. In addition, most industrial enterprises were not granted the right to undertake imports or exports by themselves. They, thus, had to rely on national or local Foreign Trade Corporations to do the actual technology imports (Grow 1987). To provide incentive for enterprises to undertake technology import and technological renovation projects in the Sixth Five-Year Plan period, various forms of preferential treatment were given to enterprises. These measures included subsidies in the form of low interest loans, grants, and foreign currency quotas, reduction of and exemption from tariffs for imported equipment and inputs, and reduction of and exemption from sales, income and adjustment taxes. Enterprises were also allowed to increase depreciation rates to speed up the renewal of fixed assets. To help finance the renovation of their own enterprises, local governments and ministries were allowed to retain 30 per cent of depreciation which had until then been handed over to the state (SEC 1985: 51–5).[23]

Administratively, technology imports into China since the early 1980s have been divided into two groups. The construction of large and new plants with imported technology is organized by the State Planning Commission and investment for these projects is made available through the basic construction plan. Technology imports to renovate existing state enterprises are managed the State Economic Commission[24] and investment for these projects is arranged through the technological renovation plan (SEC 1985: 27). The annual national plan of technology imports and technology renovation is drawn up based on the five-year plan, the industrial development programme, and the state budget. Once the budget for technology renovation has been determined, it is allocated to localities and ministries with an eye to the economic development strategy and local and ministerial interests. Localities formulate their own technology renovation plans which include projects financed by both central funding and their own funds, including foreign exchange. In turn, enterprises can also use their own retained profits and foreign exchange earnings to finance their technology import projects. Therefore, projects of technology renovation with imported technology are usually financed jointly by central, local and enterprise funds, although some projects can be financed mainly by one of these sources.

However, no matter what the financial sources are, technology renovation projects, especially the imports of technology, are subject to the approval of higher authority, according to the amount of investment, the amount of foreign exchange spending and the sources of the financial resources (SEC 1985: 299–300). Small-sized projects, especially those financed by local money, can be approved by the Planning and Economic Commissions at municipal or provincial levels, but these projects still have to be referred to the central commissions. Projects with total spending exceeding a norm,[25] no matter who is the sponsor, must be approved by the central commissions. Projects even partly financed by central funds are also subject to central approval. In order to get their projects into the national plan, localities and enterprises involved must submit a project proposal explaining the feasibility of the project. Once technology import projects are put into the national or provincial technology renovation plan, they gain certain preferential supports, such as a foreign exchange allocation and technology renovation loans.

Motivated by the preferential treatment and high profit margins, state enterprises encouraged by local governments competed to import duplicated assembly lines and equipment to earn quick returns (Conroy 1992: 212–26). The duplicated imports were mainly concentrated on assembly lines for consumer electronic goods, electrical appliances and light industrial goods. Cao Jiarui (1986) has analysed the reasons behind the problem of duplicated imports. First of all, the problem was embedded in the system organizing and implementing the programme of technological renovation with imported technology, in which the organizations in charge of technology imports were divided functionally and regionally. These organizations competed with each other for greater allocations of funds, ignoring whether they had the ability to carry out the import projects. Second, the central planners did not provide adequate information for enterprises and failed to put in place appropriate monitoring and approval procedures to control the imports by local government and enterprises. Therefore, the fundamental reason for the problem of duplicated imports was in the mandatory planning in which the subsidized funds were administratively allocated and the enterprises were eager to seek the 'rents' associated with technology import projects.

Consequently, the performance of imported technology was poor in terms of fulfilment of designed production capacity, because in some cases the importing enterprises did not possess the necessary technological capability to carry out the technology import projects and in some cases the production capacity of imported production facilities far exceeded domestic demand. Excessive imports of assembly lines also needed a large amount of foreign currency to maintain their production because these production lines relied on imported parts and components (Conroy 1992: 212–18).

Facing loosening macro control and increasing duplicated imports of production lines, the central government promulgated a series of regulations after 1985.[26] The major measure was recentralization of import approval. To tighten the macro control on equipment and machinery imports, the state stipulated in 1985 that imports of machinery using foreign exchange from central funds, no matter how large the contract was, or using local or enterprises foreign exchange more than a norm,[27] must have the permission of the Central Office of Machinery and Electronic Products Imports Approval. In the meantime, the State Economic Commission worked out several lists of imports of technology that were subject to central control. Imports of assembly lines of most electronic and electrical consumers goods, such as television sets, tape recorders, refrigerators and washing machines, and of motorcycles and automobiles were placed on the imports prohibition list. Imports of these production lines were no longer allowed. Imports of technologies and equipment which could be supplied by domestic firms were on the imports restriction list. Imports of technology, equipment and CKD kits which were on the lists no longer enjoyed preferential tariff rates. On the contrary, an import licensing system and an import adjustment tax were adopted for these imports. Under this arrangement, imports of technology, equipment and CKD kits first had to obtain approval from the central government and be brought into the national annual imports plan. An import licence would then be issued and an additional import duty (import adjustment tax) was levied. This licensing system acted as a control measure to ensure that foreign technologies imported would be in line with the industrial and technology development policy and continued to be actively used throughout the 1980s and the early years of the 1990s.[28] Technology and equipment of which duplicated imports were likely were on the unified imports list. Imports of these technologies and equipment must be planned and implemented by the relevant central industrial ministries. Imports of technology and equipment in the unified import plan no longer enjoyed duty free treatment, but they did not need to pay the import adjustment tax. In contrast to the technology and equipment on these lists, technology imports within the national and local technology renovation plans continued to enjoy the subsidized financing and tariff exemption or deduction.

Technology imports in the 1980s

As the result of the new technology import policy and the economic reforms, the technology imports of domestic enterprises changed dramatically in the 1980s (Sheng *et al.* eds 1991; Conroy 1992). Generally, imported technology was widely used in the whole economy through various channels with flexible mechanisms. Large import projects for new plants were still controlled and financed by the central commissions. However, smaller projects to renovate existing enterprises were implemented in a more decentralized way and financed by various sources of funds. These projects were closely integrated into the normal operation of industrial enterprises.

This feature of technology imports since the early 1980s has resulted in the fragmentation and incompleteness of statistics about technology imports. Furthermore, it has always been a controversial issue about what should be included in technology imports. Consequently, it is nearly impossible to give a comprehensive statistical survey on technology imports by Chinese enterprises. Nevertheless, the fragmented and limited information available can still be used to illustrate the trends of technology imports as a result of technology import policy changes and economic reforms.

As mentioned before, the focus of technology imports after the early 1980s shifted from turn-key projects to technology renovation of existing enterprises. Under this new strategy, imported technology and equipment played an increasing role in the renewal and updating of existing obsolete production facilities. Consequently, there were two approaches to the importation of foreign advanced technology throughout the 1980s. The first was the continuing importation of technology for large new plants. These projects were mainly funded by the central government and often concentrated on infrastructure facilities and on the basic industries in the petroleum, metallurgy and chemical sectors, thus reflecting the sectoral priority of central decision makers (Conroy 1992: 193). The second approach was the importation of technology in connection with existing enterprise renovation projects. Total technology imports during the period from 1979 to 1988 was reported to be US $27 billion. Among these, US $15 billion was used to import more than 20,000 items of foreign technology for existing technology renovation projects (Sheng *et al.* eds 1991: 81). This new approach emerged in 1982, when the SEC formulated the programme of 3,000 projects to renovate existing enterprises with imported technology.

There were actually 3,825 projects carried out under this programme. The total spending was US $3.68 billion, most of which (84.5 per cent) was incurred in 1984 and 1985. In contrast to the technology imports of new plants which concentrated in sectors of energy and the heavy industry, the machinery, electronics, textile and light industry sectors were given first priority in technological renovation with the imported technology programme during the 1980s. They together accounted for 75.6 per cent of the total number of contracts and 64.8 per cent of the total foreign exchange spending. The basic sectors of metallurgy, chemicals and building materials, however, accounted for only 13.0 per cent of total foreign exchange spending. The policy at that time of encouraging processing industries was responsible for the bias of investment towards the sectors of machinery, electronics, textile and light industry. The high profit margins in these sectors due to the distorted price system was another important reason for the high investment in them, which reflected the impact of the reform measure of profits retention on the technological behaviour of state enterprises and local governments.

The types of technology imported were diversified, ranging from equipment and licensing agreements to technological consulting and services. However, capital goods embodied technology, such as key pieces of equipment and production lines, still accounted for a very large share of foreign exchange spending. Some contracts were listed as licensing agreements, but the spending on 'pure' technology was still a small part of the total amount of investment. As a result, software accounted for only 12.9 per cent of the total number of contracts and 10.3 per cent of total foreign exchange spending, although the priority of government policy was on software technology such as know-how, technological documents and patents.[29]

Although it is difficult to evaluate the comprehensive performance of technology imports under the technological renovation programme, there were several major problems identified by both domestic and foreign commentators on the acquisition, utilization and assimilation of foreign technology by Chinese state enterprises (Conroy 1992; Sheng *et al.* eds 1991; and Kong and Yang 1991). The major problems in the acquisition stage of technology imports included duplicated imports of production lines and equipment and the bias of imports towards equipment and machinery. As we have seen, duplicated imports were mainly in consumer durables and downstream industries. For example, 113 TV production lines were set up in the middle 1980s. The total designed production capacity was 12.6 million sets annually, which was far more than the actual demand and led to low utilization rates and hence reduced scale economies (Conroy 1992: 216–18). These imported production lines and equipment, especially those of electronic goods, consumed a large amount of foreign exchange to maintain production. A survey of selected technology import projects in the 3,000 projects programme conducted by the Technical Economics Institute in 1987 (TEI 1988) shows that key pieces of equipment, complete sets of equipment and production lines accounted for 77.9 per cent of the total number of contracts. The accumulated foreign exchange spent on importing components and parts from 1985 to 1987 already exceeded the foreign exchange spent on importing production lines. In 1987, the ratio of foreign exchange earnings to foreign exchange investment was 17 per cent, while the ratio of foreign exchange spent on imports of production inputs to foreign exchange investment was 53.1 per cent. Problems of duplicated imports and the bias towards capital goods were common throughout the economy. In Sheng *et al.* eds (1991), there are a number of working reports on the programme of ‘technology renovation with imported technology’ by fourteen industrial ministries and forty-four localities. Among them, eight ministries and twenty-four localities reported that duplicated imports and the bias towards capital goods were serious.

The main problems in the installation and utilization stages of technology transfer in Chinese enterprises were overshooting construction schedules and a low utilization rate of production capacity caused by various factors. According to the TEI survey, 32 per cent of import projects whose installation began before 1985 had not been put into production as late as 1987. Of the projects which were put into production, 16 per cent had not reached their designed capacity. The major reasons reported by enterprises included the difficulty of obtaining components from domestic suppliers, lack of foreign currency to import inputs, poor technological capabilities, and low market demand due to duplicated imports. These problems were closely related to the poor planning and pre-investment evaluation of technology import projects and the poor management of project implementation and utilization. In the working reports in Sheng *et al.* eds (1991) mentioned earlier, a large proportion of ministries (eight out of fourteen) and localities (twenty-four out of forty-four) listed poor planning as one of the major problems in their technology import programmes (Sheng *et al.* eds 1991). Cao (1986) argues that the low utilization rate of imported technology was closely related to the duplicated imports of production lines. The major reason lay in the system governing technology imports, in which ministries and localities competed with each other for their own interests, leading to a planless and chaotic situation.

The problem in the assimilation, localization and improvement stages of technology transfer in China was mainly a lack of assimilation efforts by recipient enterprises. In the Chinese literature, assimilation is usually confused with localization. However, in this study, assimilation is defined differently from localization. The latter means using domestically supplied raw materials and intermediate inputs to replace imported ones. Localization inevitably involves assimilation, since local suppliers have to learn how to meet the specifications of the original product technology and the recipients of foreign technology need to adapt the imported technology to local conditions. Nevertheless, the assimilation of imported technology may not involve the localization of intermediate inputs. The confusion of assimilation and localization in China is probably caused by the fact that the technology import policy in the early 1980s gave priority to imports of production lines for final product assembly (Sheng *et al.* eds 1991: 88). The low local content of components naturally raised the issue of localization, and moving back along the production chain from final product assembly to the manufacture of parts and components was regarded as the measure of mastery of the imported technology. Consequently, at both the sectoral and enterprise levels, assimilation often implied extra technology import projects to allow local firms to have the capability to manufacture intermediate inputs. The 'twelve series' programme of assimilation projects is a typical example. This has led to the result that assimilation in its real meaning is often ignored, because many enterprises actually relied on imports of extra technology and equipment to localize their production of intermediate inputs.

Therefore, the assimilation efforts in this study are defined as the investment of recipients in installation, utilization, modification and improvement of the imported technology, so that the efficiency and effectiveness in transferring foreign technology can be improved. Given that a successful technology transfer needs heavy investments in learning, as we discussed in Chapter 2, deficient investment in the assimilation of imported technology was one serious problem in China's technology transfer programme. In the government's policy of technology transfer, priority was given to the purchase of technology, but not to the assimilation of imported technology. As a result, poor assimilation and lack of assimilation funds allocated through the national plan are mentioned as two of the major drawbacks in their working reports by most of the ministries and localities (Sheng *et al.* eds 1991).

In the survey by TEI (1988), 33 per cent of enterprises reported that lack of assimilation funds was one of the major problems in their technology transfer and another 39 per cent thought that their assimilation funds were not sufficient. The survey points out that one of the reasons for the lack of funds was that financial resources for assimilation were not taken into account in the government plan. However, the short term behaviour of enterprises was also responsible for this problem. For example, the government policy stipulated that enterprises could use up to 20 per cent of the profits produced by the imported technology as their assimilation funds. However, a large part of enterprises did not do that, because it would have reduced their retained profits. Even when assimilation funds were made available through the government plan, they were mostly used by enterprises to purchase capital goods, because capital goods were quick to produce profits. The result is that the increase of fixed assets was much faster than that of total assets of enterprises (TEI 1988: 25).

The shortcomings mentioned above are the manifestations of the technological behaviour of Chinese state enterprises that has ultimately been defined by government regulations, reform measures, technology import policies and market conditions. In the following chapters, how these factors and changes in these factors influenced the evolution of the technological behaviour will be analysed by using the case of China's state enterprises in the tape recorder industry. The discussions in this chapter will be used as the background information for the empirical study developed in Chapters 5 and 6.

4 Technology transfer in the Chinese tape recorder industry: the general survey

THE DEVELOPMENT OF THE CHINESE TAPE RECORDER INDUSTRY

The rapid development of the cassette tape recorder industry began in the late 1970s. Previously, tape recorders manufactured by Chinese firms were old fashioned models for professional recording of broadcasting stations. Therefore, before 1979, the number of tape recorder producers was very small and the output of tape recorders was no more than fifty thousand per year (SSB 1990: 424). In the late 1970s, cassette tape recorders began to enter the Chinese market and became very popular, especially with young people. Encouraged by the huge demand and high profit margins (mainly as a result of high tariffs on the final products at that time), the relevant industrial administrative organs and firms (mainly those previously producing radios) began to import technology to produce cassette tape recorders to substitute for imports.

The importation of technology in the tape recorder industry was a typical example of technology transfer in consumer durable manufacturing sectors in China during the 1980s. At first, the assembly lines and complete knocked down (CKD) kits were imported to produce final products. After experience was gained in assembly, the product technology, that is tape recorder design and blueprints, and the process technology and equipment were imported to localize the production of parts and components. As mentioned in Chapter 3, the approach of technology transfer starting from final product substitution was said by Chinese policy makers to have the following advantages. First of all, final product assembling with CKD kits could quickly meet domestic demand. Second, compared with imports of the final products, this approach could save foreign exchange expenditure. Third, with the high tariff, the relevant industrial administrative organs could earn high profits which could be used to push technology transfer back along the production chain. However, in practice this approach had drawbacks. The imported assembly lines needed to import large amounts of intermediate inputs to maintain suitable levels of production. Second, importation of assembly lines in itself involved little transfer of technology. Third, final product substitution protected by high tariffs together with the decentralization of decision making on technology imports caused serious duplication of imports of production lines (see Zhu Rongji 1987 and SPC 1990 in Sheng *et al.* eds 1991: 76–80 and 81–90).

The administration of the tape recorder industry

In common with all state owned industrial enterprises in China, tape recorder enterprises are administered in a cellular structure, that is to say that they are under the supervision of both the Ministry of Machine Building and Electronics Industry (MoMBEI) and the local government agencies in charge of industrial production and planning. MoMBEI normally is responsible for the coordination of production and for the long term programming and technological development of the industry. Under MoMBEI, the division specifically in charge of the tape recorder sector is the General Bureau of Communication and Broadcasting Industry (GBCBI). This division deals with all industrial development planning and investment project programming, including the planning, approval and implementation of projects of technological renovation with imported technology. The local counterparts of GBCBI of MoMBEI at provincial and municipal levels are normally called the Bureau of Electronics Industry (BEI) which is the supervisory authority for tape recorder enterprises.

At the local level, the production, basic construction investment projects and technological renovation projects of state owned tape recorder enterprises are planned, organized and implemented by the local Economic and Planning Commissions (EPC). As a result of decentralization, the local EPCs have obtained more power and resources to carry out economic activities in their own interests (Huang 1990 and Lee 1993). Administratively, the function of the local EPCs is to formulate and implement local economic policies, and to coordinate and supervise the resource allocation and project implementation of local industrial bureaux, including those of local BEI, in line with the national economic policy.

Each tape recorder enterprise is thus normally subject to the administration of a local BEI, which as the supervisory authority controls and monitors the planning and operation of enterprises. But, enterprises have to follow the policies set up by the EPC and their technological projects are subject to the approval of the EPC, on which they rely to a considerable extent for their funding. In the early 1980s, GBCBI of MoMBEI selected a number of enterprises as the appointed producers of cassette tape recorders. As appointed producers, these enterprises had their projects of technological renovation with imported technology listed in the national plan and enjoyed subsidized loans and foreign exchange.

The development of tape recorder production in China

As the result of the technology import projects carried out by tape recorder producers, the output of cassette tape recorders in China has increased very rapidly from the late 1970s (Table 4.1). There are two statistical sources of tape recorder output. The first one is the statistics of SSB (the State Statistics Bureau), which covers enterprises with independent accounting. These enterprises are normally state and urban collective enterprises, or rural collective firms at or above the township level. The second set of statistics is provided by the China Audio Industry Association (CAIA 1994). The difference between these two is that the latter includes the output of tape recorders produced by foreign direct investment (FDI) firms. In 1992, the difference between these two figures was 16.6 million sets which can be considered as mostly the output of FDI firms in Guangdong province for export through Hong Kong.

According to the SSB figures, the output of tape recorders in China was 0.17 million sets in 1979 and 30.2 million sets in 1990, growing at 60.6 per cent annually. After 1990, output began to decrease. However, the CAIA figure shows that the total output of tape recorders since 1988 has still been growing steadily, although the growth rate slowed in the last two years. This implies that FDI firms that produced tape recorders mainly for export have performed better than domestic firms in recent years.

One of the major reasons for the decline of local firms' production is the

Table 4.1 China's tape recorder output and sales volume (1979–92)

	Output of tape recorders (1,000 sets)				*Domestic sales of tape recorders*			
	SSB figures		*CAIA figures*		*Sales volume (1,000 sets)*		*Sales value*[a] *(m Yuan)*	
		Growth (%)		*Growth (%)*		*Growth (%)*		*Growth (%)*
1979	165	—	165	—	320	—	83	—
1980	743	350.30	743	350.30	1,090	240.63	392	372.29
1981	1,546	108.08	1,546	108.08	1,550	42.20	556	41.84
1982	3,471	124.51	3,471	124.51	2,570	65.81	897	61.33
1983	4,977	43.39	4,977	43.39	4,575	78.02	1,514	68.78
1984	7,764	56.00	7,764	56.00	8,435	84.37	3,003	98.35
1985	13,931	79.43	13,931	79.43	13,243	57.00	6,007	100.03
1986	17,568	26.11	19,562	40.42	15,111	14.11	6,806	13.30
1987	19,780	12.59	19,780	1.11	17,804	17.82	7,552	10.96
1988	25,404	28.43	23,441	18.51	23,909	34.29	10,142	34.30
1989	24,181	−4.81	35,633	52.01	15,375	−35.69	6,054	−40.31
1990	30,235	25.04	40,655	14.09	12,049	−21.63	4,764	−21.31
1991	28,737	−4.95	42,826	5.34	12,188	1.15	4,884	2.52
1992	27,837	−3.13	44,396	3.67	12,396	1.71	5,034	3.07

Sources: SSB, various issues, and China Audio Industry Association (1994).
[a] At current price.
Note: The sales figures are from both SSB and CAIA. The export volume of tape recorders was 13.2 million sets in 1992 (CAIA 1994).

saturation of the domestic market. In the early 1980s, the number of tape recorders possessed by per 100 households was very low.[1] The huge demand for tape recorders in the early 1980s stimulated the expansion of this industry. Because of the high tariff protection, the domestic sales volume of tape recorders grew in parallel with the output until 1988 when the domestic sales volume recorded a high of 23.9 million sets per year (Table 4.1). The expansion of domestic sales volume led to the rapid increase in the number of tape recorders possessed by households, especially in urban areas. As a result of market saturation, domestic sales volume after 1988 dropped dramatically to 12 million sets in 1990. In recent years, the domestic sales volume has remained approximately at the 1990 level.

As a result of declining domestic sales, the economic outcomes of the tape recorder industry, especially those of the state enterprises, have deteriorated in recent years. This is shown in Table 4.2, which covers only state and urban collective enterprises. It can be seen from Table 4.2 that profits and taxes per employee and the ratio of profits and taxes to the total assets of the tape recorder industry increased steadily from 1982 to reach their peak in 1985, because of the huge domestic demand and the high profit margin due to tariff protection.[2] Encouraged by this high profitability[3] and the relatively simple technology involved, local governments and enterprises competed to set up tape recorder production facilities and to import assembly lines. Consequently, supply increased much faster than demand and the market for tape recorders became a buyer's market in the second half of the 1980s.[4] Although the government imposed restrictions on the imports of tape recorder production equipment and CKD kits in 1985 to halt duplicated imports (see Chapter 3), the profitability of domestic tape recorder producers nevertheless

Table 4.2 Profitability of tape recorder producers

	Profits and taxes per capita (RMB ¥/per employee)	*Ratio of profits and taxes to total assets (%)*
1982	2,898	22.45
1983	3,052	22.27
1984	3,787	27.45
1985	4,656	32.57
1986	3,251	18.25
1987	3,376	16.85
1988	2,575	10.48
1989	1,074	4.89
1990	299	1.33
1991	313	1.32
1992	279	1.11
1993	364	1.37

Source: CAIA (1994).

fell after 1985, as the result of excess production capacity and competition from FDI enterprises. By 1990, profits and taxes per employee and the ratio of profits and taxes to total assets had dropped to RMB ¥299 and 1.33 per cent, respectively. These represented declines of 93.6 per cent and 95.9 per cent, respectively, as compared with 1985 (Table 4.2).

The development of tape deck manufacturing

As the key intermediate input of tape recorder assembly, the production of tape decks developed as tape recorder output grew (Table 4.3). However, due to the technology import strategy that technology transfer began at the final

Table 4.3 China's tape deck output (1983–92)

	Output of tape decks (1,000 sets)	*Growth (%)*
1983	3,211	
1984	4,919	53.19
1985	10,079	104.90
1986	16,703	65.72
1987	27,692	65.79
1988	37,222	34.41
1989	54,159	45.50
1990	70,485	30.14
1991	77,514	9.97
1992	91,992	18.68

Sources: SSB, various issues and CAIA (1994).
Note: It was estimated that the export volume of tape decks was 27.8 million sets in 1992 (CAIA 1994).

Table 4.4 Profitability of tape deck producers (1985–93)

	Profits and taxes per capita (RMB ¥/per employee)	*Ratio of profits and taxes to total assets (%)*
1985	7,301	37.5
1986	7,007	28.53
1987	7,579	24.02
1988	4,031	12.04
1989	2,755	9.68
1990	1,773	4.87
1991	456	1.39
1992	428	1.26
1993	468	1.35

Source: CAIA (1994).

product assembly stage, tape deck production lagged behind tape recorder assembling in the early 1980s. As is shown in Tables 4.1 and 4.3, the output of tape decks in China was smaller than that of tape recorders until 1986. The gap between demand and the domestic supply was filled by imported decks. To cut the foreign exchange expenditure through localization of intermediate input production, the state approved several tape deck technology import projects to produce tape decks locally. Again, the high profit margins (Table 4.4) due to tariff protection on imported tape decks[5] and the huge demand for this component stimulated the expansion of production capacity. As a result, the output of tape decks soon overtook the output of tape recorders. In addition, many Hong Kong producers of tape decks began to relocate their production to mainland China in the late 1980s. Consequently, the output of tape decks continued to increase in spite of the slumping tape recorder assembling business. As a result, price competition in the tape deck market has become increasingly fierce and the profitability of tape deck producers has been declining since the mid-1980s, as the output of tape decks has exceeded domestic demand (Table 4.4).

Increasing market competition in the late 1980s resulted in substantial changes in the geographical distribution of tape recorder and tape deck pro-

Table 4.5 Geographical distribution of tape recorder and tape deck production in China

	Tape Recorders				*Tape Decks*			
	Producers		*Output*		*Producers*		*Output*	
	Number	*%*	*1,000 sets*	*%*	*Number*	*%*	*1,000 sets*	*%*
1987								
Beijing	8	1.8	870.3	4.4	2	4.1	1,350	4.9
Tianjin	6	1.4	138.5	0.7	4	8.2	3,954	14.3
Shanghai	64	14.4	3,283.5	16.6	6	12.2	2,563	9.3
Jiangsu	72	16.3	4,371.4	22.1	12	24.5	9,825	35.5
Zhejiang	38	8.6	1,068.1	5.4	n.a.	n.a.	n.a.	n.a.
Guangdong	137	30.9	6,092.3	30.8	13	26.5	5,175	18.7
Others	118	26.6	3,956.1	20.0	12	24.5	4,825	17.4
Total	443	100.0	19,780.3	100.0	49	100.0	27,692	100.0
1992								
Beijing	4	1.0	596.8	1.3	1	1.6	1,021	1.1
Tianjin	4	1.0	612.3	1.3	2	3.2	3,510	3.8
Shanghai	23	6.0	1,633.1	3.6	4	6.5	2,005	2.2
Jiangsu	42	11.0	8,505.7	18.8	9	14.5	9,210	10.0
Zhejiang	22	5.8	1,955.2	4.3	n.a.	n.a.	n.a.	n.a.
Guangdong	188	49.2	27,078.8	59.7	29	46.8	61,996	67.4
Others	99	25.9	4,981.7	11.0	17	27.4	14,250	15.5
Total	382	100.0	45,363.6	100.0	62	100.0	91,992	100.0

Source: CAIA (1994).

duction and in the type of ownership of tape recorder producers (Tables 4.5 and 4.6). The production technology of tape recorders and tape decks was imported first by state enterprises in North and East China in the early 1980s. This is because state enterprises had privileged access to subsidized financial resources and foreign exchange, both of which were strictly allocated through state planning. Moreover, the areas of North China and East China were given priority in the early 1980s (see Chapter 3). As a result, in the mid-1980s, the tape recorder industry was dominated by state enterprises in these areas. In 1987, these areas produced more than 50 per cent of the total output of tape recorders and more than 60 per cent of the total output of tape decks. Moreover, state enterprises in these areas were technologically more advanced and enterprises in other areas obtained their technology mainly from them.

After five years of development, the patterns of both the geographical distribution of tape recorder production and the type of ownership of tape recorder producers changed considerably. Guangdong Province, where FDI firms are concentrated, outpaced North China and East China in tape recorder output growth and became the largest producing region in China. In 1992, its output of tape recorders and tape decks accounted for 50.3 per cent and 67.4 per cent, respectively, of the total output of China (Tables 4.5 and 4.6). During the period 1987–92, the share of state enterprises in the total number of tape recorder producers decreased dramatically, while the share of collective and FDI firms increased rapidly. This shows that these non-state firms were more competitive than state enterprises and implies that they were profitable, although the economic performance of most state enterprises was poor (Tables 4.2 and 4.4).

BACKGROUND OF THE SAMPLE ENTERPRISES

In order to conduct the questionnaire interview, thirty-one enterprises were selected as the sample. The names of the sample firms are given in Appendix 1. The criteria for choosing these thirty-one enterprises can be summarized as follows: (1) the sample firms are mainly state owned and quasi state owned enterprises that are located in the major production centres of tape recorders

Table 4.6 Types of ownership of tape recorder producers in China

	1987		*1992*	
	Number	*%*	*Number*	*%*
State owned	245	55.2	149	39.1
Collectively owned	185	41.8	206	53.9
FDI	13	3.0	27	7.0
Total	443	100.0	382	100.0

Source: CAIA (1994).

in North China and East China; (2) they have conducted technology transfer projects and have been producing tape recorder products continuously since then; and (3) they were cooperative and gave easy access to information.

Among these thirty-one sample enterprises, twenty-three are in the tape recorder assembly sector, and the other eight enterprises are tape deck producers. The average proportion of tape recorder related sales to total sales of these firms was around 70 per cent in 1993. Although the tape recorder sales of some enterprises only account for less than 20 per cent currently, they were mainly manufacturers of tape recorders during the 1980s and had conducted technology import projects involving tape recorder production. Due to the sluggish tape recorder market, they subsequently changed their product mix and mainly reverted back to the products they had manufactured before the technology of tape recorders was imported. A large proportion of the sample tape deck producers assembled tape recorders as well, but this study is only concerned with their technological activities in tape deck production.

From these sample enterprises, four enterprises were chosen for in-depth case studies in Chapter 7. Of these, two are in the tape recorder assembling sector and two are tape deck producers. The tape recorder firms are Canzhou Recorder Factory (CRF) and Westlake Recorder Factory (Westlake). The case study samples of tape deck makers are Weihai Recorder Factory (WRF) and Wuxian Wireless Factory (WWF).

All but two of these thirty-one sample enterprises are state owned and urban collective enterprises. The exceptions are two township enterprises (see Chapter 5 for details of one of these township enterprises).[6] The urban collective firms in China are nominally owned by all members of the firms. However, in reality, the government treats them more or less the same as state enterprises, because of their blurred ownership. Historically, these urban collective enterprises were mainly formed by merging several small privately owned workshops and radio repair shops during the socialist transformation movement in the mid-1950s.[7] Since then, these firms have been merged and dismantled several times and have undertaken several investment projects financed by local governments. As a result, a large part of the assets of these firms has been contributed by the government, although they still remain collectively owned. To a large extent, the government supervises the urban collective firms in a manner similar to the state enterprises. In practice, these firms enjoy lower government subsidies in investment, because they are only in the lowest level of planning and rarely given grants from state or provincial governments. This may explain why there are fewer collective firms in tape deck production, which needs relatively heavy investment.

All these sample firms are medium-sized and small firms under the supervision of local governments.[8] Although their supervisory governmental agencies are called variously the Electronics Industrial Bureau, the Instruments-making Industrial Bureau, the Electronics Industrial Office, the Communication and Broadcasting Equipment Company, etc., depending on the organizational set up of different localities, these agencies are responsible for the operation and development of the local household electronics industry. This implies that the influence of local government on the technology transfer process in the sample firms is larger than the influence of state technology transfer policy. As state and quasi state enterprises, all of these firms experienced the enterprise reforms in the 1980s, such as the tax-for-profit, loan-for-grant, and CMRS schemes (see Chapter 3). Nearly all of them adopted the Contract Management Responsibility System. The starting time of the first run of CMRS of these firms ranged from December 1986 to May 1988. The termination date of the contract for all firms was in 1990, which was the last year of the Seventh Five-Year Plan (1986–90). The governmental contractors were normally the supervisory agency (for example, the Electronics Industry Bureau), the Economic Commission, the Finance Bureau, the Taxation Bureau, and the local branch of the Industrial and Commercial Bank. Sometimes, the CMRS contract was signed between the firm and the supervisory agency or the administrative holding company. In this case, the administrative holding company (representing all firms under its supervision) signed an overall contract with the Economic Commission, Finance Bureau, Taxation Bureau and the Bank. During the second run of the CMRS which started in 1991, the State Assets Administrative Bureau, which is a newly established governmental agency to prevent the state assets from being eroded, joined the ranks of governmental contractors.

On the side of the enterprises, the signatory person was normally the managing director who, since the introduction of CMRS, has become the legitimate person in charge of the enterprises, replacing the communist party secretary. However, as these remained state enterprises, the appointment of the director was basically made by the supervisory agency and the Economic Commission. In the case of large firms, the appointment had to go through the cadre selection procedure of the personnel department of the party.[9] In some enterprises, competent people were allowed to apply and compete for the position of the director, although the appointment had to be approved by the higher authority. In the samples, twenty-four enterprises had their directors appointed by the government. The directors of the other seven firms (including the two TVEs) had been selected through the combination of direct election and government approval.

Because these firms are in the sector of consumer goods production, their technological capabilities are relatively weak, as compared with large plants in heavy industry. Among the thirty-one sample firms, there was only one enterprise (Yangcheng Wireless Factory) that established its own tape recorder R&D department. Other enterprises had a technical department which was responsible for all technological matters, mainly those in production. In the sample enterprises, the percentage of technological personnel to total employees was higher in state owned enterprises than in collective and township enterprises. On average, the figure is 10 per cent in state owned enterprises and 6 per cent in non-state firms (own interview information).

Most of our sample firms had some prior experience in making electronic products. A large proportion of these firms (nineteen out of thirty-one) were established in the 1950s through merging several private firms. One firm was set up in the 1930s. The remainder were relatively new. Twenty-one out of twenty-three tape recorder producers used to make wireless products. Most of them were semi-conductor radio producers. Of the eight tape deck producers, only two had not produced wireless products prior to production of tape decks. However, they had experience in making metal parts.

Relatively speaking, the sample firms are more capital intensive than the industry average. The average per capita fixed assets of the sample tape recorder firms was 11.5 per cent more than the whole sector and the figure of sample tape deck producers was 7.9 per cent higher than the industry average in 1993 (Table 4.7). This can be explained by the fact that state enterprises account for a larger proportion of the sample firms than of the industry as a whole.

Influenced by the slumping domestic market for tape recorders, the economic performance of the sample enterprises was poor in recent years (Table 4.7). In 1993, the ratio of profits and taxes to the total capital of the sample firms was 2.07 per cent for tape recorder firms and 1.19 per cent for tape deck producers, similar to the figure for the industry as a whole. In the sample firms, there were nine tape recorder firms and two tape deck producers that were suffering losses. However, non-state firms performed slightly better than state enterprises, although the latter were relatively well-equipped. Of the eleven loss makers, nine are state owned, and the other two are urban collectives. Of our sample tape recorder producers, there are five firms whose ratio of profits and taxes to total capital was higher than 6 per cent in 1993. Three out of these five firms were non-state firms.

Table 4.7 The comparison between the sample enterprises and the tape recorder industry (1993)

	Fixed assets per employee (RMB ¥/ person)	*Total capital per employee (RMB ¥/ person)*	*Sales per employee (RMB ¥/ person)*	*Profits and taxes per employee (RMB ¥/ person)*	*Ratio of profits and taxes to total capital (%)*
Tape recorder					
total industry	13,564	26,567	21,539	364	1.37
sample firms	15,121	31,804	42,265	506	2.07
Tape deck					
total industry	21,643	34,787	38,255	468	1.35
sample firms	23,344	50,243	44,678	598	1.19

Source: CAIA (1994) and own interviews.

TECHNOLOGY IMPORTS OF THE SAMPLE ENTERPRISES

The technology of tape recorder production was mainly imported by Chinese state enterprises, first through assembling imported CKD kits and then through purchasing foreign equipment and blueprints to localize the production of components and parts. In this investigation, the focus is on the latter, because in the first stage the purpose was mainly to meet the increasing domestic demand and little technology was transferred. The technology import projects undertaken by the sample enterprises were mainly conducted during 1984 and 1985, the last two years of the Sixth Five-Year Plan (1981–5) when the 3,000 projects programme was planned and implemented. Only five enterprises carried out their technology import projects after 1985, because in June 1985 the state promulgated a new policy to cut down on duplicated imports of tape recorder assembly lines and related equipment.

In the late 1980s, the technology transfer process of the sample firms entered a phase which was defined as the utilization stage in the analysis framework (see Figure 2.1). During this stage, many sample firms had conducted additional technology import projects. Because the purpose of these projects was to use the originally imported technology better or to further localize parts production, they are discussed in detail in the following sections concerning the assimilation efforts of the sample firms.

The production process of cassette tape recorders and tape decks

The production process of cassette tape recorder in a typical Chinese tape recorder producing enterprise consists of the major production activities including: (1) final product assembling; (2) electric parts soldering and assembling; (3) plastic parts manufacturing; and (4) mould/die making (see Appendix 2). From the beginning of the tape recorder production process, the task of mould/die making is to make moulds used as accessory tools for the extruders to produce plastic parts. Plastic parts manufacturing includes the press moulding of plastic parts and surface mounting of plastic parts. The soldering and assembling of electrical parts include the soldering of electric circuit boards and the assembling of outside sourced electrical parts and, components, such as tape decks, speakers and switches, with the electric circuit board to make up the electrical part of a cassette tape recorder. The final stage is to put the assembled electrical parts, outside sourced components and plastic parts in the outer case, using an assembly line.

Most tape recorder assemblers do not produce tape decks by themselves, because of the different technology involved and the need for the economies of scale. However, all tape deck producers also assemble tape recorders. A tape deck consists of metal parts, plastic parts, a mini-motor, magnetic heads and other electrical components. Tape deck manufacturers use outside sourced mini-motors, magnetic heads and electrical components. However, they normally produce most of the metal parts and plastic parts by themselves. The typical production process of Chinese tape deck factories includes: (1) mould/die making; (2) the production of plastic parts; (3) the production of punched metal parts and processed metal parts; and (4) the assembly of tape decks (see Appendix 2).

The major technologies and equipment imported by tape recorder enterprises

In Chinese government documents and relevant statistics, technology imports are normally classified as complete sets of equipment, key pieces of equipment, licensing agreements, technology consultation and technical services.[10] As we mentioned in Chapter 3, in the programme of renovating existing enterprises with imported technology, state enterprises at first were encouraged to import key pieces of equipment, assembly lines and disembodied technology, that is to make licensing agreements and purchase technology consultation and services, in order to re-equip and upgrade their obsolete production facilities. Later on, the central planners began to restrict assembly lines and certain types of equipment due to the problem of duplicated imports of capital goods. Therefore, the technology import projects of the sample firms during the technology acquisition stage mainly involved the imports of key pieces of equipment and assembly lines. Several of the sample firms also concluded licensing agreements with foreign firms. Nevertheless, the number of firms which licensed foreign technology only accounted for a small proportion of the thirty-one sample firms. Moreover, the foreign exchange spent on licensing technology was only a small fraction of the total investment (see discussions later on). During the late 1980s, some of the sample firms conducted additional technology import projects to localize the production of parts. As we will see later, the technology they imported was also mainly capital goods.

As noted, most of the sample tape recorder producers were radio manufacturers before the tape recorder product technology was imported by them. Therefore, the technology involved in electrical parts assembling and plastic production was not entirely new to them. However, the equipment of these firms was obsolete and of poor quality. Therefore, most sample firms emphasized the importation of process technology, equipment and production lines. The major equipment imported to update the process technology of electrical parts assembling included automatic electronic component connectors, electronic audio signal generators, and wave soldering equipment to speed up the assembling and to improve the quality of assembled electric parts. As far as plastic parts production was concerned, the equipment imported by the sample firms was usually mould extruders and plastic part coating lines.

Tape decks for cassette tape recorders were a totally new product to Chinese enterprises in the late 1970s. The product technology was imported in the early 1980s by several Chinese enterprises after a few years of assembling tape decks with CKD kits on the imported assembly lines (see case studies in Chapter 6). Relatively speaking, tape deck assembling is simple, involving little transfer of technology. However, the associated product technology and especially the process technology are complicated, because the requirements for precision and durability of parts are quite high. Chinese tape deck producers often lacked the necessary designing and processing experience for metal parts manufacturing. Therefore, the technology imported in the first half of the 1980s consisted mainly of processing equipment and techniques for parts processing, especially metal parts processing and punching, as well as the blueprints of tape decks. Later on, some firms also modernized their design technology by importing computer aided design (CAD) technology.

In order to improve the quality and increase the efficiency of metal parts processing, the sample tape deck firms mainly imported high speed and multiposition automatic lathes and die casting presses. As far as the punching of levers and brackets is concerned, the major technology and equipment imported included multipoint punches and stepping dies. Multipoint punches could produce several pieces of parts at the same time. The stepping dies could finish several processes of punching and folding of a punched part at one stroke, so that not only was it possible to increase the speed of processing, but the quality of the finished parts was improved greatly.

For both tape recorder assemblers and tape deck producers, mould/die making was a particularly weak aspect, because they had not specialized in metal parts processing. The problems were mainly (1) obsolete designs, (2) poor fitting techniques, and (3) out-dated machine tools and measurement instruments. To improve their technological capabilities, some sample firms imported the dies and moulds which they later copied. Some firms imported the design technology and assembling and adjusting know-how to make dies and moulds by themselves. In addition, they usually imported specialized machine tools to replace the old ones and sent technicians abroad to learn how to operate these machines and fit the dies. The imported die/mould making equipment mainly comprised high precision machine tools and measuring instruments, such as numerically controlled millers, thread cutters, spark cutters, and rangefinders.

PERFORMANCE OF THE TECHNOLOGY IMPORT PROJECTS

As was discussed in Chapter 2, an effective technology transfer at the enterprise level can be defined as the improvement of economic performance caused by the increase of technological capability of the recipient firms. Therefore, the performance of imported technology utilization by the sample enterprises is measured in terms of their increase in technological capabilities as a result of the technology import projects and their overall economic performance in using the imported technology.

Technological capability improvement

The effect of technology transfer projects on the technological capability improvement of recipient enterprises can be reflected in the recipient's ability to implement and use the imported technology, and to improve and innovate on the basis of the learned technology. In terms of the analytical framework of the study (see Chapter 2), the former is viewed as the result of the first stage of assimilating imported technology and the latter is that of the advanced stage of assimilation.

In the investigation, it was found that all firms reported that they had put the imported technology into production and a large part of the sample enterprises thought that they had basically mastered the imported technology and their products could meet the technical specification of the foreign technology (Table 4.8). They also said that they had the ability to maintain the imported facilities by themselves, although several firms reported that the breakdown rate of some imported facilities was quite high, due to lack of spare parts and inexperience in operation. When asked about the efficiency in

Table 4.8 Technological capability of recipients in implementing and using imported technology

	Responses	*% of total responses*
1. Implementation time		
a. ahead of scheduled time	0	0
b. on schedule	19	61.3
c. behind scheduled time	12	38.7
2. Mastery of the imported technology		
a. fully mastered	6	19.4
b. basically mastered	24	77.4
c. not really mastered	1	3.2
3. Utilization rate as compared with the planned capacity		
a. larger than 100%	13	41.9
b. 80 to 100%	12	38.7
c. 50 to 80%	4	12.9
d. less than 50%	2	6.5
4. Technological specification		
a. fully met	4	12.9
b. basically met	27	87.1
c. not completely met	0	0
d. not met at all	0	0
5. Maintenance of imported equipment		
a. by the firm itself	9	29
b. by firm itself except major problems	22	71
c. reliance on foreign firms	0	0

Source: Own interviews.

implementing and using the imported technology, many enterprises admitted that their performance was not satisfactory. Nearly 40 per cent of sample enterprises could not meet the time schedule for the installation of imported technology. More than half of them have never fulfilled the planned output set in their feasibility studies.

Table 4.9 lists the technology activities the recipient firms have carried out

Table 4.9 Technological capability of recipients in adapting and modifying imported technology

	Responses	*% of total responses*
1. Adapting imported technology to local conditions		
a. yes, independently	6	19.3
b. yes, assisted by local organizations	18	58.1
c. yes, assisted by foreign suppliers	3	9.7
d. no	4	12.9
2. If yes, significance of the modification		
a. very significant	0	0
b. significant	12	44.5
c. not very significant	11	40.7
d. minor	4	14.8
3. New product development as the result of technology transfer project		
a. yes, independently	6	19.4
b. yes, assisted by local organizations	19	61.3
c. yes, assisted by foreign suppliers	6	19.4
d. no	0	0
4. Imitation of imported equipment		
a. yes, independently	5	16.1
b. yes, assisted by local organizations	8	19.4
c. no	18	64.5
5. Innovation based on imported technology		
a. yes	1	3.2
b. no	30	96.8
6. Fostering local suppliers of parts and components		
a. yes	24	77.4
b. no	7	22.6
7. Transferring imported technology to other local firms		
a. yes	5	16.1
b. no	26	83.9
8. Localization rate of imported products[a]		
a. 100%	3	9.7
b. 80 to 100%	28	90.3
c. 50 to 80%	0	0
d. less than 50%	0	0

Source: Own interviews.

[a] In terms of the value of indigenously made inputs.

to improve the imported technology. These reflect the enhancement of technological capability of the sample firms as the result of their technology import projects. In the questionnaire interviews, more than three-quarters of the sample firms reported that they did some kind of modification to the imported technology to adapt it to local conditions. No firm thought that their modification was very significant, although half of the firms which had undertaken an adaptation said that the change was significant. Further enquiry found that the change in imported technology was mainly in order to use locally available raw materials and components.

All of the sample tape recorder enterprises reported that they developed some kind of new products as the result of technology transfer projects. One-fifth of the firms carried out their new product development tasks independently. Others did so with the assistance of other firms. However, further investigation revealed that the development of new products by the sample enterprises was limited, in terms of both novelty and speed of new product development. Many firms only copied what had been done by other firms and the lead time of their new product development was lengthy. Consequently, they often missed out on opportunities in the marketplace. For example, one of the sample firms decided in 1986 to launch a cassette recorder/radio with dual decks, a model which was in short supply at that time because young people used it to duplicate pop songs. It took more than half a year for the firm to develop this new product. When the firm finally began to assemble the new product in 1987, dual deck recorders were over supplied and prices had declined sharply.

Because tape recorder firms are not manufacturers of capital goods, they by-and-large did not undertake activities of reverse engineering. More than one-third of the sample firms reported that they imitated imported facilities. However, they mainly meant that they had copied the imported assembly lines and die/moulds. One firm reported that it worked together with a machine building enterprise to imitate the imported assembly line and sold several sets of the copied assembly lines to other domestic firms. Nevertheless, the majority of the sample firms did not imitate the imported equipment. They were basically passive users of the purchased capital goods.

As far as the diffusion of imported technology is concerned, the picture is mixed. Most of the sample firms fostered local suppliers to secure sources of components. In the investigation, it was found that many firms helped their small suppliers to produce certain kinds of components special to tape recorders or tape decks, such as non-standardized springs and washers. They also relocated some kinds of production activities, such as heat treatment and electroplating, to rural areas, mainly due to environment protection regulations. However, most firms tried to keep the major production activities within their own firms. Only five firms reported that they transferred the imported technology to other firms. This might be due to the fact that most of the sample firms imported capital goods rather than licensed technological know-how. Another possible explanation is that the market mechanism for domestic technology transfer has not yet been established.

Localization of the production of imported product is one of the major objectives set out in the government's technology transfer policy.[11] In this respect, the performance of the sample firms seems satisfactory at first glance. None of the sample firms had a localization rate lower than 80 per cent. A few firms even had all parts and components made in China. However, further investigation showed that, although most factories made parts by themselves, they still relied on imports for some inputs, such as high quality steel sheets, coating powders and high-strength plastics. Further, the localization rate of some high grade products, such as high fidelity stereo systems and walkmen was relatively low.

As was discussed in Chapter 2, the highest level of technology transfer and the most difficult stage in the assimilation process is innovation based on the mastery of foreign technology. In this respect, most of the sample firms did not advance to this stage. In the investigation, only one firm said that it had made an innovative change in its imported technology. The other thirty firms reported that their R&D activities focused mainly on localization of parts production and designing new products by following the technology development trend in the world market for tape recorders.

From the figures listed above, two conclusions may be drawn. First, Chinese tape recorder makers were generally successful in installing and utilizing imported technology, especially capital goods embodied technology, in their production. They were also able to produce tape recorders for which the product technology was imported to the required specification and quality standard. In order to use domestic materials/parts/components, a number of these firms modified the imported technology, both in terms of product design and processing technology, to adapt it to local conditions. On the other hand, they were normally just passive users of the imported technology. They basically remained at the primary stage of technology transfer. As will be discussed in the following sections, their efforts to master the imported technology mainly concentrated on technological activities at the lowest level of assimilation, that is the utilization of imported technology. Consequently, few firms have ever reached the higher stages of the technological capability ladder.

Second, the efficiency of the sample firms is poor, in terms of both their economic performance in using the imported technology in production and their technological capability improvement as compared with the resources and time invested. As shown in the investigation (see discussions below), a large proportion of tape recorder firms have never met their profitability target set in their feasibility studies of technology import projects. Nevertheless, in the early stage of tape recorder production in China in the early 1980s, they had enjoyed on average a profitability higher than the average of the whole industrial sector, mainly because demand was much higher than supply and as state enterprises they were the first to acquire the imported production facilities.

Economic performance

The major indicators used to measure the sample enterprises' economic performance are output of tape recorder products, total profits and taxes, and gross value of industrial output (GVIO).[12] The reason for choosing these standards is that they are common indicators of enterprise performance in China and available for most of the sample enterprises. In this study, the actual performances of these three indicators in the first and third years after the project was completed, and in 1993, are compared with the planned targets in the feasibility study of individual technology import projects. In this comparison, the planned targets are used mainly as a frame of reference to show the change of their overall economic performance over the period from the mid-1980s, when the projects started, to the present.

Generally, the sample enterprises, most of which conducted their technology projects during the period from 1983 to 1985, have experienced a rise and fall in the last decade similar to that of the whole tape recorder industry. When we evaluate the overall performance of the sample firms, this general trend of market development in the tape recorder industry should be borne in mind. Tables 4.10 and 4.11 illustrate the change in the overall economic performance of the sample tape recorder producers and of tape deck firms, respectively, over time.

The overall performance of technology import projects in the sample firms was not satisfactory when compared to the targets set in their feasibility studies. As shown in Tables 4.10 and 4.11, Chinese tape recorder technology importing firms on average achieved nearly 80 per cent of their output and GVIO targets in the first year. After two years of improvement, their average output and GVIO reached more than 100 per cent of the targets. Entering the 1990s, however, the average output and GVIO of these firms showed no improvement. As a group, they managed to maintain their output targets as compared with the planned ones. However, this is mainly because some firms expanded their production capacity in the second half of the 1980s and as a result produced more than the output target set in the original feasibility studies. As a result of price decreases as the market for tape recorders became saturated, the average GVIO of sample firms, in particular the tape recorder producers, dropped to even less than the targets of the mid-1980s.

If measured in terms of profits and taxes, the overall performance of the sample firms is even less satisfactory than in terms of output and GVIO. On average, they never reached the planned profits and taxes target. In the first three years after the project completion, that is approximately the period from 1985 to 1988 which was the best time for the tape recorder industry in China, the average profits and taxes were only 40 to 60 per cent of the target for tape recorder producers and 50 to 80 per cent for tape deck firms. In 1993, the sample firms on average only produced around 10 per cent of their targeted profits and taxes, meaning that the production costs of these firms increased during this period.

Table 4.10 Performance of technology import projects of tape recorder producers

	First year	*Third year*	*Current (1993)*
Output as compared with planned target			
Number of firms whose output is			
larger than 100%	3	10	8
80 to 100%	5	2	1
50 to 80%	6	3	2
30 to 50%	2	2	2
less than 30%	1	0	4
Average of 17 firms, % of target	78.6%	105.9%	97.3%
GVIO [a] *as compared with planned target*			
Number of firms whose GVIO is			
larger than 100%	4	8	4
80 to 100%	4	5	1
50 to 80%	6	2	5
30 to 50%	2	2	0
less than 30%	1	0	7
Average of 17 firms, % of target	79.9%	102.5%	74.6%
P&T [b] *as compared with planned target*			
Number of firms whose P&T is			
larger than 100%	1	0	0
80 to 100%	2	2	0
50 to 80%	5	6	0
30 to 50%	4	0	2
less than 30%	2	5	8
minus	0	1	4
Average of 14 firms, % of target	64.2%	43.6%	10.1%

Source: Own interviews.
[a] GVIO stands for Gross Value of Industrial Output which is similar to turnover but at constant prices.
[b] P&T stands for Profits and Taxes. They are in current prices.
Note: Detailed information for some firms is not available. Seventeen firms provided data about their output and GVIO. Fourteen provided P&T data.

The change of overall economic performance of sample firms can also be shown by the number of firms whose performance, judged by these three indicators, reached a certain percentage of their targets. It can be seen from the tables that the third year after project completion was the best time in terms of output and GVIO. A large part of the sample firms achieved more than 100 per cent of their targets of output and GVIO. However, in 1993, the number of firms that produced more than their targeted output and GVIO declined. Many firms produced less than half of their targets. Again, if measured in terms of profits and taxes, the number of firms that have ever reached their targets is very small. Only a few firms fulfilled their profits and taxes target in the first years after project completion (one out of fourteen tape recorder producers during the first year and two out of six tape deck

Table 4.11 Performance of technology import projects of tape deck producers

	First year	*Third year*	*Current (1993)*
Output as compared with planned target			
Number of firms whose output is			
larger than 100%	1	4	2
80 to 100%	0	0	1
50 to 80%	4	2	2
30 to 50%	1	0	1
less than 30%	0	0	0
Average of 6 firms, % of target	74.7%	126.7%	106.8%
GVIO[a] *as compared with planned target*			
Number of firms whose GVIO is			
larger than 100%	1	4	1
80 to 100%	0	0	1
50 to 80%	5	2	1
30 to 50%	0	0	3
less than 30%	0	0	0
Average of 6 firms, % of target	80.2%	124.4%	94.7%
P&T[b] *as compared with planned target*			
Number of firms whose P&T is			
larger than 100%	0	2	0
80 to 100%	1	2	0
50 to 80%	2	0	0
30 to 50%	1	1	1
less than 30%	2	0	3
minus	0	1	2
Average of 6 firms, % of target	49.5%	80.6%	8.7%

Source: Own interviews.
[a] GVIQ stands for Gross Value of Industrial Output which is similar to turnover but at constant prices.
[b] P&T stands for Profits and Taxes. They are in current prices.
Note: Six out of eight tape deck firms provided their output, GVIO and P&T data.

firms during the third year). In the 1990s, as the performance of the tape recorder industry deteriorated, the profits and taxes produced by most of the sample firms dropped to less than 30 per cent as compared to the target of their feasibility studies. A large proportion of the sample firms even incurred losses.

5 The technological behaviour of Chinese SOEs: the general survey

THE TECHNOLOGICAL BEHAVIOUR OF SOEs

As was discussed in Chapter 3, economic reforms in China during the past fifteen years have dramatically changed the institutional environment under which state owned enterprises (SOEs) operate. This change has been reflected in the technological behaviour of SOEs in importing and utilizing foreign technology and has affected negatively the improvement of their technological capabilities, because the partially reformed institutional arrangement did not provide incentives for or pressures on SOEs to make long term efforts in technological activities. In particular, the institutional arrangement characterized by nonexclusive property rights to the assets of industrial production did not encourage SOEs to choose appropriate types of technology, to use the imported technology efficiently and to make adequate assimilation efforts to build indigenous technological capabilities. Specifically, the technological behaviour of the sample SOEs in technology transfer can be summarized as a propensity to use capital goods embodied technology and a lack of assimilation efforts in learning imported technology.

The propensity to use capital goods

The investigation found that the sample SOEs tended to carry out projects which mainly involved imports of capital goods. The propensity to invest in capital goods in the technology acquisition period provided additional evidence for the hypothesis of the traditional SOEs behaving like an 'engineering man' and a 'bureaucratic man' in importing foreign technology (see Chapter 2 of this study and James ed., 1989). This kind of technological behaviour is a manifestation of the agency problem caused by the hierarchical administration of SOEs and can be characterized as the propensity of SOEs to carry out as many large projects as possible to import capital intensive technology.

Nevertheless, the preference of the sample SOEs for capital goods was different from the technological behaviour of state firms in the pre-reform period, because as a result of economic reforms, the objective of SOEs as a collection of their total employees changed from pursuing an easy life to seeking their own income maximization (Walder 1989, Lee 1991 and Wu Jinglian 1993a). Specifically to the sample firms, they preferred to import capital goods mainly because (1) their projects were financed and subsidized by the government through administrative channels; (2) the use of capital goods could generate a quick return in a protected market; and (3) they were allowed to retain a proportion of the profits generated from the investment.

During the late 1980s when the CMRS system was implemented by SOEs, the sample firms entered the utilization stage of technology transfer. Under this institutional arrangement of CMRS, the property rights structure of SOEs became further blurred and the common property problem inherent in SOEs became increasingly serious. SOEs tended to increase the short term cash flow of their lifetime employees by sacrificing the long term development of enterprises (see Chapter 2). Under this arrangement, if they could still obtain subsidized financial resources from the government and if they had to make an investment of their own funds in technological activities, they preferred to use capital goods which could produce quick returns.

The acquisition of foreign technology during the mid-1980s

Table 5.1 gives the types of technology import projects undertaken by the thirty-one sample firms during the mid-1980s. These projects were classified by the major types of technology imported, as shown in the first column in this table. The second column gives the minor types of technology imported by the sample firms, because some projects involved more than one type of technology. The investigation found that, although the government policy on technology transfer during the mid-1980s encouraged state enterprises to import disembodied technology, such as product designs, production know-how and technical consultation and services (see Chapter 3), the technology import projects of the sample firms concentrated on imports of capital goods. As shown in Table 5.1, all of these projects involved imports of capital goods. The most popular type of technology imported was key pieces of

Table 5.1 Types of technology imported by sample enterprises

	Major type	*Minor type*
Key piece of equipment	24	4
Complete set of equipment	1	0
Assembly line	5	0
Technology licensing agreements	1	4
Others[a]	0	3
Total	31	11

Source: Own interviews.
[a] 'Others' include technology consultation and services.

equipment. Among these thirty-one projects, twenty-four imported mainly key pieces of equipment. Another four firms imported key pieces of equipment together with other types of technology. The number of firms which imported entire assembly lines was much fewer than the number that imported key pieces of equipment. In fact, only one firm imported complete sets of equipment as their major type of imported technology. The reasons are that the import of complete sets of equipment was excluded in the programme of renovating existing enterprises with imported technology and that government policy during 1984/85 began to discourage the imports of assembly lines. Consequently, projects involving imports of these two types of technology were difficult to get approved by the authorities.

Among the sample enterprises, five firms licensed foreign technology. Of these, two licensing contracts were in the tape recorder assembling sector. Both centred on auxiliary technology related to the decoration of plastic parts and assembled tape recorders. The other three licensing contracts were in the tape deck sector. Two of these involved the core technology of tape deck product designs, while the remaining one was about auxiliary technology related to die making and designing. The figures in Table 5.1 show that the attitude of Chinese firms towards technology transfer was that they preferred capital goods to licensed technology. Moreover, they seldom bought foreign technology in the form of technology consultation and services.

The bias towards capital goods can also be found through the actual investment spending on different types of technology. From Table 5.2, it can be seen that on average the cost of technology imports accounted for 61.6 per cent of the total actual investment of technology import projects of the sample enterprises. The remaining 38.4 per cent of the total investment was implementation costs. Taking the technology import costs as a whole, the cost of purchasing capital goods accounted for 83.3 per cent, licensing agreements for 4.1 per cent, production inputs for 7.6 per cent, and the 'others' for 5 per cent. Production inputs normally included raw materials and components to test the imported machinery. The 'others' covered the costs of technical consultation and services of foreign suppliers, and the training of a recipient's employees abroad or at home. Since nearly every project involved some kind

Table 5.2 Investment on different types of technology

Total actual investment (%):	100.0
Technology import costs	61.6
Implementation costs	38.4
Total technology import costs (%):	100.0
Capital goods	83.3
Licensing agreements	4.1
Production inputs	7.6
Others	5.0

Source: Own interviews.

of training abroad, technology services purchased from foreign suppliers accounted for only a very small fraction of the total costs of imports.

Total implementation costs of these technology import projects were less than two-thirds of total technology import costs. The implementation costs included the construction and re-building of workshops (34.2 per cent), installation and trial operation of imported equipment (12.4 per cent), translation, compilation and sorting of foreign technology documents and blueprints (6.7 per cent), purchase and installation of auxiliary machinery and instruments (26.9 per cent), trial production (10.6 per cent) and training (9.2 per cent). Implementation means the installation and utilization of the imported technology in the production process. Although it is a part of the assimilation process, our sample firms' implementation investment mainly consisted of installation costs.

Additional technology imports during the late 1980s

In the late 1980s, the sample firms entered the utilization stage of technology transfer during which some sample firms continued to carry out additional technology import projects. Although these projects were undertaken by the sample firms in the name of localization of parts production (see Chapter 4), the real purpose was mainly to expand output, because under the CMRS system, the sample SOEs were allowed to retain a large part of their incremental profits. Consequently, these additional technology import projects were biased towards purchase of foreign equipment which could be put into production quickly.

Interview information shows that among the thirty-one sample enterprises, there were fourteen firms which carried out additional technology import projects. Of these fourteen firms, only two concluded licensing agreements with foreign suppliers. Significantly, for the other twelve firms, additional technology imports only included the purchase of capital goods and production inputs from abroad. The bias towards capital goods in additional technology imports can also be found in the actual spending of foreign exchange on different items. As shown in Table 5.3, for all sample firms whose statistics are available, equipment accounts for the largest part of their foreign exchange spending. The average percentage of equipment in total foreign exchange spending was larger than that in the original technology import projects. On the other hand, as they were mainly intended to localize parts production, imports of inputs accounted for less than previously.

The lack of learning efforts to assimilate imported technology

As mentioned earlier, the process of technology transfer of the sample SOEs entered the utilization and assimilation phase during the late 1980s when the CMRS system was implemented by SOEs. Under this arrangement, SOEs were required to make technological investment with their retained profits

Table 5.3 Foreign exchange spendings of extra technology import projects

Firms	*Total investment m RMB ¥*	*Foreign exchange m US $*	*Equipment % in FE*	*Licensing % in FE*	*Inputs % in FE*	*Others*[a] *% in FE*
BWIF	9.31	1.16	73.2	12.4	0	14.4
BCWF	2.20	0.64	100.0	0	0	0
T731	3.22	0.47	75.4	24.6	0	0
N3RF	2.02	0.24	88.2	0	0	11.8
NDWF	3.77	0.85	90.6	0	0	9.4
NDMF	1.95	0.21	100.0	0	0	0
NREF	2.61	0.38	100.0	0	0	0
YWF	3.85	0.57	100.0	0	0	0
CRF	10.61	2.17	89.2	0	5.2	5.6
NHWF	3.51	0.51	96.5	0	0	3.5
BHRF	6.56	0.84	91.2	0	4.5	4.3
WRF	3.14	0.76	100.0	0	0	0
N8434	2.27	0.41	85.4	0	8.6	6

Source: Own interviews. FE stands for foreign exchange.
[a] 'Others' include training abroad and technological services of foreign firms.
Note: See Appendix 1 for the full names of the sample firms.

(see Chapter 3). However, the investigation found that the sample firms tended to invest less than they should have in the assimilation of imported technology. Specifically, when their products were profitable, they invested in expansion of output. When their profits declined, investment in technological activities was the first area to be cut. (See the case studies in Chapters 6 and 7.)

The field research identified four broad categories of activities undertaken by Chinese tape recorder firms to make better use of imported technology (Table 5.4). 'Additional technology imports' refers to the importation of foreign equipment and licensing agreements with foreign firms to supplement the original imported technology. 'R&D' covers technological projects to improve both the imported product and process technologies. 'Training' refers to the training of local employees in foreign countries or at home by foreign suppliers and other training programmes specifically relevant to the implementation and utilization of the imported technology. 'General management' mainly includes total quality control (TQC), and the change of organizational structure and the application of new management techniques appropriate to the imported technology. The last type of assimilation activity was normally carried out in the course of the daily operation of the recipient firms. Therefore, although twelve firms reported that they had undertaken this type of assimilation activity, mainly TQC and changes in working procedures, detailed information about this is not available for the most part.

As mentioned above, additional technology imports of the sample firms involved mainly the purchase of equipment aiming at expansion of output.

Table 5.4 Assimilation activities undertaken by the sample tape recorder firms

	No. of firms who conducted the activity	*No. of firms who did not conduct the activity*
1. Additional imports of technology	14	17
2. R&D projects	17	14
3. Training	31	0
4. Improvement of management	12	19

Source: Own interviews.

Therefore, they are not considered as genuine assimilation efforts. Among the thirty-one sample firms, there are seventeen firms that carried out R&D projects. Because many of them undertook both additional technology import projects and R&D projects, eleven firms (which account for 35.5 per cent of total sample enterprises) undertook neither additional technology imports nor R&D projects. In other words, additional technology import projects and/or R&D projects were restricted to only twenty firms. These projects were largely implemented during the period from 1984 to 1989 when enterprise reform in China proceeded into the stages of tax-for-profit and CMRS and the market demand for tape recorders was rising. After 1989, there were only two firms which carried out major R&D projects. All firms claimed that they had conducted training programmes and new product development activities. As we will see later, there were several firms which tried to diversify into the production of CD players and VCR players through technology imports. However, these are new technology transfer projects, rather than assimilation activities for the original projects of imported tape recorder technology.

The results of the enterprise interviews reveal that the sample firms generally had not made sufficient R&D investment. According to the stipulation of government, state enterprises could use as high as 1 per cent of their total sales on their R&D activities. This expenditure could be treated as production costs. In the investigation, it was found that the R&D expenditure of most firms between 1989 to 1993 was very small, accounting for only 0.3 to 0.5 per cent of their total sales. The R&D spending of a few firms was relatively large. However, this was mostly financed by bank loans.

Some firms gave the reasons for their failure to spend more on R&D activities. First of all, in recent years, their profits had been shrinking. They did not have enough financial resources to invest in R&D activities. Relatedly, a few firms said that they were planning to enter the CD player or video recorder businesses, but they could not get bank loans to carry out their technological renovation projects. Second, the contracts of the CMRS required firms to hand over a fixed amount of profits (see Chapter 3). When the R&D spending of firms increased, their retained profits would be reduced, even though their R&D expenditures could be recovered through sales. Third, R&D was a long term orientated activity whose future outcome was uncertain. In addition, it would use a large amount of funds which were in seriously short supply for tape recorder firms. Using more funds on R&D inevitably would have aggravated their difficulties in current operations.

To facilitate the technological renovation of existing enterprises with imported technology, the central government encouraged state enterprises to organize training programmes. Training costs were allowed to be counted as production costs. The costs of training abroad and inviting foreign experts to China could be listed in the foreign exchange budgets of technology import projects (SEC 1985: 55 and 364). Consequently, most firms undertaking technology import projects had sent their people to be trained abroad. On average, training costs account for 3 to 5 per cent of total foreign exchange spending for tape recorder firms and 5 to 7 per cent for tape deck firms. Complete data about how many people had been trained abroad are not available. However, according to a few firms' records, management personnel accounted for a large proportion of total employees trained abroad. According to discussions between the author and interviewees, training in a foreign country was viewed as a great benefit to the trainees, because they were allowed to bring back duty free electronic goods. Tape deck firms seem to have sent more machine operators as well as more employees abroad than tape recorder assemblers, probably because the equipment imported by tape deck firms included more sophisticated machine tools. Nevertheless, it was found in the investigation that training in general was not given high priority in most of the sample firms. In some firms, 50 per cent or more of workers did not receive any form of formal training. In recent years, faced with deteriorating profits, the sample firms have spent less and less money on training. According to government regulations, firms were allowed to spend as much as 2.5 per cent of their total wage bill on their training programmes. However, most sample firms used only about 1 per cent of their total wage bill on training. The major reason was that within the limited budget, the priority was given to the welfare of their employees.

MACROECONOMIC FACTORS AND THE TECHNOLOGICAL BEHAVIOUR OF SOEs

From the above figures, the technological behaviour of the sample tape recorder firms in transferring foreign technology can be characterized as having had a bias towards capital goods and by a lack of assimilation efforts, which led to the unsatisfactory performance of technology transfer projects in the tape recorder industry. What, then, are the reasons that state enterprises failed to make sufficient assimilation efforts to improve their performance in using the imported technology?

In order to answer this question, in the questionnaire, a series of questions is included under two headings: (1) the positive factors for firms to carry out assimilation projects; and (2) the negative factors discouraging firms from undertaking assimilation projects. The sample firms were asked to rank the possible factors listed in the questionnaire. The ranks are on a scale from 1 to 5, with 5 denoting the most important variable and 1 the least important one. The answers given by the sample firms are then summed up and the average ratings of each possible variable ranked from high to low, as listed in Tables 5.5 and 5.6 respectively.

The data in these two tables and other interview information show that the technological behaviour of the sample SOEs in transferring foreign technology was ultimately defined by the macroeconomic factors of the institutional arrangements governing their operations, the government policy directing their technological import projects, and the market environment in which they competed with each other and with firms with other types of ownership. Interview information also shows that other factors, such as the micro technical factors played a relatively small role in determining the technological behaviour of the sample firms. The questionnaire interview information showing the impact of macroeconomic factors is thus tabulated and discussed under the headings of economic reforms, government policy, and market competition.

Table 5.5 Positive factors for Chinese firms in making assimilation efforts

Variables	*Ratings*
1. The availability of financial resources (bank loans for technology renovation)	4.52
2. Growth of market demand for the product	4.45
3. The high profit margin for updated quality goods	4.10
4. The good economic performance of your firm	4.05
5. Market pressures from competitors and customers	3.95
6. The appropriateness of previously imported technology	3.82
7. Government sponsorship of the assimilation project of your firm	3.16
8. Government preferential treatment for assimilation activities	2.68
9. The requirements of the government	2.65
10. Technical assistance from domestic firms	2.36
11. The technological capability of your firm to carry out assimilation projects	2.26
12. The requirement of the CMRS contract	2.14
13. CMRS allows firm to retain more funds for assimilation	1.96
14. Technical assistance from domestic S&T[a] institutions	1.87
15. Technical assistance from foreign suppliers	1.76

Source: Own interviews.
[a] S&T stands for science and technology.
Note: The number of respondents is 31. The rating is on a scale from 1 to 5, with 5 representing the most important factor and 1 the least important one.

Table 5.6 Negative factors for Chinese firms in making assimilation efforts

Variables	*Ratings*
1. Lack of finance resources	4.77
2. Competition from imported goods	4.65
3. Assimilation costs too high	4.48
4. The short term effects of CMRS	4.25
5. The current enterprise administration system	4.10
6. Assimilation efforts too slow to be effective	4.03
7. Duplicated imports of similar technology that led to over-supply	3.85
8. Inadequate preferential treatment from government to assimilation of imported technology	3.45
9. Imported technology not appropriate	3.25
10. Lack of technical assistance from domestic firms	3.10
11. Government restrictions on investment projects	2.94
12. Lack of technical assistance from domestic S&T[a] institutions	2.61
13. Price controls for new products	2.58
14. Lack of technical assistance from foreign suppliers	2.55
15. Restrictions from foreign suppliers	1.39

Source: Own interviews.
[a] S&T stands for science and technology.
Note: The number of responses is 31. The rating is on a scale from 1 to 5, with 5 representing the most important factor and 1 the least important one.

Economic reforms and the technological behaviour of SOEs

As was discussed earlier, economic reforms in China during the past fifteen years were carried out through a responsibility system within the framework designed for central planning (Jefferson and Rawski 1994, Wu Jinglian 1993a, and Lee 1991). Consequently, the decision making autonomy of SOEs was enlarged and the property rights structure of SOEs were further blurred. Under this institutional arrangement, the technological behaviour of SOEs (characterized by the propensity for using capital goods and the lack of learning efforts) can be attributed to the agency problems and common property problem caused by the ambiguous property rights in state owned enterprises under a hierarchical structure of central planning.

Agency problem

The process of technology transfer of the sample firms proceeded alongside the economic reforms. They undertook their technology import projects during the mid-1980s when enterprise reform was at the stage of profit sharing (the tax-for-profit period) (see Chapter 3). During this period, the financial resources were still allocated to state enterprises through the state plan, but the SOEs were given more autonomy than before in decision making regarding the specific projects and were allowed to retain a proportion of the profits generated from the investment. Because the objective of the government as the principal was unlikely to be identical with that of SOEs (as the collection of their management and employees) as the agent, when the 'management' of SOEs was separated from the 'ownership' within the hierarchical framework designed for central planning, the agency problem became serious (James ed. 1989). Specifically, SOEs tended to use capital goods such as assembly lines and key pieces of equipment (as shown in previous section), as long as they could put their projects into the plan because the costs of capital to them were relatively low and market pressure was weak due to government protection. Therefore, changes in the degree of decision autonomy and in the objectives of the sample firms at the different stages of their technology transfer process are analysed to show the institutional reason for the agency problem, because this problem is closely related to the scope available for managerial discretion and the objectives of the agents (James ed. 1989).

During the mid-1980s when the sample firms carried out their technological import projects, they were still mostly financed by government funds. However, their autonomy in decision making regarding technology import projects was increased as the result of economic reforms. In the interviews, two questions were asked concerning who was the major initiator of technology import projects and who was the major decision maker in selecting specific technology and foreign suppliers. The answers of interviewees are listed in Table 5.7. It can be seen that local governments in the mid-1980s played a key role in the technology imports of the tape recorder industry, while central ministries delegated their decision making power to localities. Moreover, enterprises had gained an increasing weight in proposing investment projects. That local governmental agencies were the major initiators of the technology import projects of the sample firms was mainly due to two

Table 5.7 Decision maker of technology acquisition in the mid-1980s

	1. Who was the major initiator of the technology import project of your firm?		*2. Who was the major decision maker in choosing foreign technology and supplier?*	
	Responses	*%*	*Responses*	*%*
Central ministries	1	3.2	0	0.0
Local government	10	32.3	4	12.9
Local government consulting firm's opinion	14	45.2	15	48.4
The firm itself	6	19.4	12	38.7
Total responses	31	100.0	31	100.0

Source: Own interviews.

factors. First, in the administration of the programme for technology renovation with imported technology, the decision making power was delegated to, and the available funds were sliced up and distributed to, localities. Second, tape recorder firms were relatively small and under the jurisdiction of local governments.

On the other hand, enterprises had more autonomy in selecting specific technology and foreign suppliers. Only four firms reported that the technology decision was exclusively made by the government. This reflects that, although local governments controlled the power to decide the general direction of investment, they gave more autonomy to firms in specific technology decisions, because (1) firms had the necessary concrete knowledge and (2) the investment was in the form of bank loans which firms bore the obligation to pay back. Many enterprises reported that they participated in the whole process of technology importation, including searching for foreign suppliers, negotiations, on-the-spot investigation of foreign firms, and the signing of the contract.

In spite of the increasing decision making power of enterprises, all enterprises said that they had to obtain approval from various government departments and that the application process was lengthy and troublesome.[1] Many firms attributed the delay in project completion to the complicated application and approval process.

The investigation results show that the enterprises obtained more decision making autonomy in their assimilation projects than in their original technology import projects. Table 5.8 shows that none of the assimilation projects of the sample firms was initiated and decided entirely by the government. Most firms had the autonomy to initiate and determine their specific projects. However, enterprises complained that their investment decision making autonomy was still small, because they had to get approval from many government agencies, even though the projects were financed out of their own

Table 5.8 Decision making autonomy of SOEs in assimilation projects in the late 1980s

	Responses	*%*
1. Project was planned, financed and determined by government	0	0.0
2. Project was initiated by firm and financed by and subject to government plan	4	20.0
3. Project was initiated and financed by firm, but subject to government approval	11	55.0
4. Project was financed and decided by firm and reported to government agencies	5	25.0
Total number of responses	20	100.0

Source: Own interviews.

funds. In addition, firms did not have the right to deal with foreign suppliers directly. On the other hand, five firms said that their investment decision autonomy was much larger than in the early 1980s.

During the mid-1980s, the purpose of the sample firms in acquiring foreign technology was to put their projects into the state plan and import advanced machinery. Technology in the thinking of Chinese firms and government departments was equivalent to codified knowledge embodied in capital goods, technical documents and blueprints. When choosing foreign technology, engineering considerations dominated and costs were given lower priority. When asked what was the major criterion in choosing an imported technology, more than half of the sample firms answered that it was the degree of technological advancedness. Furthermore, many factories imported key pieces of equipment with much more capacity, speed and automation capability than was actually needed. For example, one firm (SEAF) imported wave soldering equipment whose capacity was five times as large as the total assembly capacity of the firm. The machine had to be used every working day, but only for two hours. To preheat the soldering tin, a large amount of electric power was wasted, in addition to the extra purchase cost of the machinery.

To put their projects in the state plan was also important for the sample firms. It was found in the investigation that there were four projects designated as 'technological diagnosis', because projects under this title would be easier to get approved. In fact, the actual imports of all these four projects were mainly capital goods. This was the major reason that many firms imported key pieces of equipment, because key pieces of equipment met both the stipulation of government policy and the objective of SOEs to get subsidized funds.

This 'engineering man' and 'bureaucratic man' behaviour was further reinforced by the incentive to obtain a proportion of profits generated from the investment. The production of tape recorders was a lucrative business during the mid-1980s when the demand was much larger than the supply and the domestic market was highly protected from foreign competitors. Relatively cheap capital costs and a high degree of managerial discretion, due to the management of state owned firms through a hierarchical structure, provided an opportunity for managers of SOEs to achieve their goals by using capital goods. In the feasibility studies of the projects, profits were the major target of the sample firms. The high profits target also reflected the bureaucratic behaviour of the sample firms in obtaining the support from local governments which wanted to expand their revenue base. In the sample firms, there were twenty firms which were financed by the renovation bank loans through the plans of local governments. This was the major reason for the duplicated imports of production facilities in the tape recorder industry.

Tables 5.9 and 5.10 list the goals of concrete additional technology import projects and R&D projects undertaken by the sample firms during the utiliza-

Table 5.9 Goals of extra technology import projects

	First goal		*Second goal*	
	Responses	*%*	*Responses*	*%*
To localize parts production	2	14.3	3	27.3
To increase output	8	57.1	2	18.2
To improve quality	2	14.3	4	36.4
To balance production	2	14.3	2	18.2
Total responses	14	100.0	11	100.0

Source: Own interviews.
Note: The number of firms which conducted additional technology import projects is 14. Only 11 firms gave their second goal.

Table 5.10 Goals of R&D projects

	First goal		*Second goal*	
	Responses	*%*	*Responses*	*%*
To develop new product	7	41.2	0	0.0
To increase output	5	29.4	6	50.0
To localize parts production	5	29.4	3	25.0
To improve quality	0	0.0	1	8.3
To cut production costs	0	0.0	2	16.7
Total responses	17	100.0	12	100.0

Source: Own interviews.
Note: The number of firms who conducted R&D projects is 17. Only 12 firms gave their second goal.

tion stage of their technology transfer. It was found that the major goals of firms were: (1) to increase their output; (2) to balance their production capacity; (3) to develop new products; and (4) to localize their parts production. Only two firms said that their objective was to improve the quality of their products by importing measuring instruments. Two firms reported that their second goal of R&D projects was to cut production costs.

That the goals of the sample firms in additional technology import projects and R&D projects were mainly focused on output expansion and parts production capacity is not surprising, because these projects were largely launched in 1985–8 when the tape recorder market was booming. After their original technology import projects, the sample tape recorder firms established their assembly capacity and some of them also acquired the capability to produce some of the parts. However, the production capacity of most of them was not balanced and many firms were still relying to a degree on imported intermediate inputs. Since the firms could not always obtain foreign exchange, however, they were no longer able to use imported parts and components to assemble the final products. As a result, many firms tried to build up their own parts production facilities. That is the major reason that a large proportion of the assimilation projects aimed at localizing parts production or at increasing output by expanding or balancing their parts production capacity. Therefore, the assimilation projects of most sample firms were in fact to complete the original projects. In other words, they were mainly at the lower level of assimilation, that is utilization of imported technology, rather than at the higher level of assimilation, that is improvement of imported technology and innovation.

Furthermore, in this period, as profits became the most important measure of enterprise performance and were related increasingly closely to the interests of enterprises and their employees, the assimilation projects became more profit oriented. This is another reason why the additional technology import projects and R&D projects of the sample firms aimed at output expansion, because the demand for tape recorders seemed to be growing rapidly. The motive for obtaining more profits is shown in Table 5.11. The majority of the sample firms that undertook additional technology import projects and R&D activities stated that profitability was the most important criterion for them to select alternatives. However, there were only two firms that said that the total cost of projects was their selection criterion. This is contradictory to the finding (see below) that lack of financial resources was the major negative factor for firms to carry out assimilation projects. A possible explanation might be that state owned tape recorder firms at that time still had access to the subsidized bank loans and that tape recorder production was a lucrative business. Once firms obtained the necessary funds, their major concern was the level of profits they could generate from the investment rather than how much they could save by selecting a cheaper alternative.

The data above show that the decision making autonomy of the sample firms in technological activities was enlarged during the reform period. As

Table 5.11 Selection criteria of alternatives for additional technology import and R&D projects

	Responses	*%*
1. Rate of return on investment	14	70.0
2. Total cost of project	2	10.0
3. Possibility to get allocated funds from government	0	0.0
4. Technological advancedness	4	20.0
5. In line with government policy	0	0.0
Total responses	20	100.0

Source: Own interviews.
Note: The number of firms which conducted additional technology import and R&D projects is 20.

was discussed in Chapter 3, the enlargement was achieved unevenly within the hierarchical system designed for central planning. In the early and mid-1980s, the investment of SOEs was still strictly controlled and financed by the government. However, managers of SOEs were given limited discretion in the selection of specific projects, as long as their projects were brought into line with the state plan. Because the technology import projects of the sample tape recorder firms were mainly in the programme of renovating existing enterprises with imported technology, they enjoyed highly preferential treatment and larger autonomy in carrying out these projects (see Chapter 3 and the discussion below about government policy). Due to the weak pressure from the government as the principal inherent in the hierarchical structure of state ownership (James 1989: 6), the sample firms had more chance to pursue their own objectives.

Common property problem

Economic reforms in the past fifteen years characterized by the responsibility system actually gave partial property rights to SOEs. They had the rights to decide how to use the assets and the rights to claim a part of the income generated from the assets. They were required to take the responsibility for their own performance and to invest a stipulated part of their retained profits in technological activities. However, the property rights were neither clearly defined nor exclusive (see Chapter 3). The assets of SOEs were still owned by the state. Consequently, SOEs as a collection of their lifetime employees tended to maximize their near term income and not to take the responsibility when they incurred losses. Specifically, they tended to increase their current income by cutting investment in technological development which could produce profits only in the future. This behaviour together with their incompetence in competing with firms with other types of ownership led to a sharp decline of profits of SOEs. (The role of market competition is discussed below. The impact of different regulations for SOEs and firms in the non-state sector is discussed in Chapter 7). Poor economic performance further reduced the financial resources which could be used in technology development. Consequently, many sample firms lost their technological capabilities established through the previous technology import projects. This shows that the common property problem brought on by ambiguous property rights caused the deterioration of technological capabilities and the dissipation of rents generated from this valuable resource.

Separation of the 'management' from the 'ownership' of SOEs through the responsibility system within the hierarchical structure of a command economy was the major characteristic of the economic reforms in China during the past fifteen years (Wu Jinglian 1993a; Tam 1992; Jefferson and Rawski 1994). In order to find out the impact of the CMRS system on the technological behaviour of the sample firms, the CMRS is listed as both a favourable and an unfavourable factor affecting the decision of the sample firms in technological activities. The results show that many firms thought the CMRS was an important negative factor for them in making assimilation efforts (Tables 5.5 and 5.6). Many firms pointed out that although firms had the obligation to fulfil technological renovation tasks as specified in the contract, the constraint to fulfil technological tasks was the weakest, as compared with the targets of profit remittance and employees' welfare. They thought that the CMRS made them focus on the achievement of short term goals, while R&D projects were mainly long term in orientation. Furthermore, many firms complained that the CMRS was the major reason for them to be short of funds, because the profit remittance quota was determined in 1987/1988 when tape recorders were lucrative products. (See the case study of CRF in Chapter 6). A few firms said that the CMRS influenced their assimilation efforts positively, because it allowed them to retain more profits. However, the number of these firms is smaller than the number which thought CMRS was a negative factor. As a whole, the sample firms thought that the current enterprise administration system was an important negative factor in regard to their operations in general and to technological efforts in particular. They thought the current system governing the operations of SOEs was far less favourable than that for FDI firms or TVEs.[2] They said that this was one of the major reasons that they could not compete with non-state firms.

That firms were short term orientated under the CMRS system can also be inferred from the high ratings of the two negative factors in Table 5.6 that assimilation was too costly and that assimilation efforts took too long to be effective. This result shows that under the CMRS system, the priority of firms was on the short term targets of the contract. The investment in assimilation could only produce profits in the future and would reduce their income at present. Therefore, these two factors discouraged their investment in technological activities.

The poor economic performance of sample firms after the late 1980s due to the incompetence of SOEs in market competition had a negative impact on technological capability buildup and improvement. In this situation, the technological capability of firms not only did not improve, but actually deteriorated. This is because technology as an intangible industrial asset depreciates fast. If firms do not keep investing resources in improving and maintaining it, it will soon lose its value. When firms performed poorly, they did not have sufficient resources to invest in technological activities. Furthermore, technology is usually embodied in capital goods, in human beings and in firms as organizations. Until the early 1990s, there was no mechanism in China for effectively relocating the production factors of poorly performing state firms. As a result, if a firm was forced out of business or to do other technologically unrelated business, the technological capability embodied in the firm would be wasted.

In the investigation, it was found that a few tape recorder firms lost at least a part of their technological capability built in the past as a result of poor performance. Table 5.12 shows that some of the sample tape recorder firms lost a large fraction of their technical personnel during the period from 1990 to 1993 when tape recorder firms were undergoing a difficult time. According to the information obtained from the questionnaire interviews, the thirteen tape recorder firms together lost 581 technical people between 1990 and 1993, which accounted for 2.1 per cent of their total employees. More importantly, the loss of technicians in some firms was quite dramatic. For example, seven firms lost more than 25 per cent of their technicians after 1990. These firms were mainly located in large cities. The performance of firms in large cities was particularly bad due to various reasons which will be discussed in the case studies. Moreover, there were more alternative opportunities to earn a higher income for qualified and experienced technicians in large cities.

The investigation found that, due to the personnel reform, state enterprise employees gained much more freedom and a greater chance to choose their jobs. The popular choices were jobs in trading companies or in FDI firms. For more experienced technicians, working at a TVE as an executive director could pay ten times as much as working in their original state enterprises. As shown in Table 5.13, doing trading business regardless of whether it was related to their technological experiences or working for FDI firms or TVEs with their technological knowledge were the two major reasons for the technicians leaving these ten sample firms.

In China, state enterprises have the function to provide secure income and welfare for their lifetime employees. Until recently, state enterprises were not

Table 5.12 Loss of technical personnel of selected firms

	1990			*1993*			*Loss of technicians 1990–1993*	
		Technicians			*Technicians*			
Firms	*Total employee number*	*Number*	*As % of total*	*Total employee number*	*Number*	*As % of total*	*Number*	*%*
BWIF	1,824	224	12.3	1,658	162	9.8	–62	–27.7
BCWF	1,521	108	7.1	1,132	55	4.9	–53	–49.1
YWF	4,331	416	9.6	4,234	428	10.1	12	2.9
CRF	3,295	388	11.8	3,013	337	11.2	–51	–13.1
CAEF	1,689	62	3.7	1,704	70	4.1	8	12.9
YWGF	1,946	187	9.6	1,949	191	9.8	4	2.1
S22W	3,169	393	12.4	2,445	249	10.2	–144	–36.6
SH742	2,586	296	11.4	2,115	229	10.8	–67	–22.6
Westlake	652	72	11.0	460	33	7.1	–39	–54.2
BHRF	815	89	10.9	785	64	8.2	–25	–28.1
WRF	1,216	141	11.6	929	82	8.8	–59	–41.8
WWF	3,216	311	9.7	2,542	225	8.9	–86	–27.7
SHWF	1,029	106	10.3	896	87	9.7	–19	–17.9
Total	27,289	2,793	10.2	23,862	2,212	9.3	–581	–20.8

Source: Own interviews.

Table 5.13 Estimates of leaving technicians' destination (%)[a]

Firms	*FDI or TVEs*	*Xia Hai*[b]	*Retirement*	*Others*[c]
BWIF	50	30	20	0
BCWF	30	40	15	15
CRF	40	40	20	0
YWGF	30	30	10	30
S22W	40	40	20	0
SH742	40	40	10	10
Westlake[d]	30	30	30	10
BHRF	30	30	10	30
WRF	30	30	20	20
WWF	40	30	10	20

Source: Own interviews.

[a] These are the estimates given by the interviewees when asked what jobs the leaving technicians were most likely looking for.

[b] 'Xia Hai' means doing business, in particular trading business.

[c] This includes transferring to government organizations, working at other state enterprises, etc.

[d] Westlake has an early retirement scheme due to poor performance. See the case study of Westlake for details.

allowed to sack their redundant employees (see Chapter 3). Because the young and well-educated and trained employees could find jobs elsewhere and most of the aged or less capable people could not, the burden of old state firms was very heavy. Furthermore, under the current state enterprise administration system, merging with other firms or going into bankruptcy were seldom solutions for loss-making firms. In addition, local governments either had no financial ability or were not willing to bail out all state firms facing a difficult situation. The major measure adopted to 'save' these firms was to divide them into smaller accounting units and let them do whatever they thought appropriate to self-rescue, as shown by the evidence in Table 5.14.

In order to earn incomes to feed their employees, many firms diversified into other businesses (Table 5.14). The interview information shows that some firms had gone back to the original products that they made before the technology import project for tape recorders, because they still had certain technical advantages in the old products. A few other firms tried to develop new products which were technologically related to tape recorders in one way or another. For example, several sample firms said they had, or would have, additional technology import projects to produce CD players. The major difficulty for them was to obtain the necessary funds. Nevertheless, the majority of sample firms reported that they had other businesses basically not related to the imported technology to provide income earning opportunities for their employees. Opening retail shops and restaurants, renovating their plant buildings into hotels, office space or warehouses to earn rents, and organizing small service businesses were among the examples. There were two reasons for them to do so. First, they were obliged to provide jobs for their

Table 5.14 Changes in business structure of some tape recorder firms

	1989			1993					
Firms	Total turnover m RMB ¥	TR&TD[a] turnover m RMB ¥	% of TR&TD to total	Total turnover m RMB ¥	TR&TD turnover m RMB ¥	% of TR&TD to total	% of TRNP[b] to total	% of OP[c] to total	% of TUB[d] to total
BWIF	67.29	49.32	73.3	44.65	13.93	31.2	9.3	25.5	34
BCWF	54.84	47.82	87.2	38.07	14.47	38	7.6	16.1	38.3
TBWF	41.01	36.62	89.3	35.39	14.23	40.2	0	35.2	24.6
NDMF	49.80	47.36	95.1	48.83	35.21	72.1	0	27.9	0
S22W	104.31	69.26	66.4	42.20	5.11	12.1	12.5	39.1	36.3
S13W	70.27	44.41	63.2	32.54	5.67	19.1	16.5	30	34.4
SAEF	12.50	10.71	85.7	10.84	2.44	22.5	13.2	40	24.3
SH742	58.25	55.34	95	45.61	6.07	13.3	0	64.4	22.3
SRAF	53.84	47.22	87.7	67.99	24.62	36.2	63.8	0	0
Westlake	38.22	38.22	100	4.03	2.87	71.2	0	0	28.8
WRF	53.95	47.21	87.5	29.94	16.2	54.1	0	30.5	15.4
XRF	51.95	49.35	95	46.27	32.16	69.5	0	30.5	0

Source: Own interviews.

[a] TR stands for tape recorders and TD for tape decks.

[b] TRNP means technologically related new products. All enterprises reporting that they have TRNP, except SRAF and BCWF, have just begun to assemble CD players. SRAF's new product is VCRs. BCWF's new product is small computer game players.

[c] OP means original product before tape recorder. Original products of tape recorder assemblers mainly include wireless equipment, such as transceiver broadcasting equipment, and radios. Of the two tape deck producers, WRF's original products are measuring and cutting tools and XRF's original products are relays.

[d] TUB means technologically unrelated business. It includes various businesses, ranging from technical services, buying and selling audio equipment, to hotels, restaurants and retail shops selling various consumer goods.

lifetime employees. Second, they did not have enough technological or financial resources to enter a new business which could better use the existing facilities. However, many of the sample firms were located in or near downtown areas and they in reality had the right to use their premises. Therefore, doing these kinds of businesses became the natural choice of these firms. In order to conduct other lines of business, many firms had closed their assembly lines and removed their facilities. For example, BCWF emptied one of its major buildings to renovate it as a hotel. It had eight assembly lines in the mid-1980s, but in 1993 only three of them were still in existence.

In the investigation, it was found that more than 80 per cent of sample firms had carried out various sideline businesses in tertiary industry, to solve the problem of providing income earning opportunities for their workers when their production of tape recorders had declined since the 1980s. For some enterprises, the income from sideline businesses accounted for a large proportion of their total incomes (Table 5.14). Doing technologically unrelated businesses might provide incomes for the employees of these firms. Nevertheless, it meant that the technological capability embodied in the firms as an organization and in human beings gained as a result of the technology import project was no longer used, and therefore was being lost.

Most of the sample firms have reduced their production of tape recorders or tape decks. As a result, the imported technology and capital goods are no longer utilized as fully. Although complete data on the utilization rate of imported equipment are not available, some cases reported by sample firms imply that a low utilization rate was common for them. For example, one firm imported two pieces of high precision 3-D distance measuring instruments that were used in its die making department. The utilization rate was always quite low in the past. In recent years, however, the firm had its dies made by Hong Kong firms, because the cost was lower than if they made the dies by themselves. As a result, the 3-D measuring instruments and many other high precision machine tools had become totally redundant. In addition, from the figures in Tables 5.10 and 5.11, it can be seen that the output of half of the sample firms was less than 80 per cent of their planned targets ten years ago. Thus it can reasonably be inferred that a large percentage of imported capital goods were not fully utilized.

Under the current institutional arrangements, the assets of the loss making firms cannot readily be re-allocated to other firms requiring similar technical equipment or expertise. As a result, the technological capability of sample firms established in the past has not been maintained, because of the loss of technicians and of switching into other businesses which are unrelated to the imported technology.

Government policy towards technology imports

In China, the government policy regarding industrial development was mainly for the state sector, especially before the mid-1980s, because the non-state sector was relatively insignificant at that time. Moreover, because the government was both the owner of SOEs and the policy maker, the instruments of industrial policy in China were closely related to the governance and control of the government over its subordinate enterprises. Furthermore, the behaviour of SOEs was ultimately defined by the basic institutional arrangement of state ownership. Therefore, the response of SOEs to the government policy towards technology transfer was shaped by and subject to the basic institutional arrangement of state ownership.

As was discussed in Chapter 3, the government policy regarding technology changed in the early 1980s from constructing large turn-key projects to renovating the existing enterprises with imported technology. In the machine building and electronics sectors, which included the tape recorder industry, the strategy of the central planners was to import the product technology and re-equip the existing enterprises and to substitute local production for the imported goods in the final stages of the production chain. This strategy as compared with the previous strategy of emphasizing turn-key projects had some advantages (see Chapter 3). Nevertheless, this policy of import substitution triggered serious duplicated imports of assembly lines and key pieces of equipment. This is because the subsidized financial resources made available through the programme of renovating existing enterprises were administratively allocated to localities and enterprises which were delegated decision making autonomy in the planning and implementation of specific projects. Although the central planners stipulated that priority should be given to software technology, the imports of capital goods was much more than licensed technology and technical services.

In order to control the duplicated imports of capital goods, the central government tightened the approval of project proposals and used administrative measures such as import licences to stop the imports of certain types of equipment. At the same time, localization of the production of parts was encouraged. Projects aiming at import substitution were given priority in financial resource allocation (see Chapter 3). In addition, the central government formulated certain policy measures to encourage enterprises to invest more in training and R&D. However, the response of SOEs to the government policy was a continuing bias towards capital goods and insufficient investment in building up indigenous technological capabilities. This is because the government relied on intervention into the operations of firms to implement its policy, rather than on regulating the behaviour of firms. Therefore, firms in the bureaucratic hierarchy of state ownership, where monitoring and control were inherently ineffective, tried to manipulate and take advantage of the policy. In the following part of this section, the role of government policy in shaping the technological behaviour of the sample firms is analysed together with the programme of renovating existing enterprises with imported technology in the mid-1980s and the localization programme in the late 1980s. In addition, the impact of government requirement on technological decisions of the sample firms is discussed here.

Nearly all of the technology import projects of sample enterprises were carried out within the category of technology renovation. This means these projects were handled by the Economic Commission, and the preferential policy for technology renovation with imported technology had applied to these projects before the new policy of June 1985. In the sample enterprises, twenty-six firms were the appointed producers of tape recorders selected by the government,[3] implying that their technology import projects were in the plan. The remaining non-appointed producers were all in the non-state sector. Once the projects were brought into the plan, they would have been allocated subsidized technology renovation loans and foreign exchange at the official exchange rate.[4] The interview information reveals that the major financial source of the technology import projects was the technology renovation loans, which accounted for 73 per cent of the total investment for the appointed producers. Firms also obtained government grants which accounted for 13.1 per cent of total investment. Only 13.9 per cent of the investment was financed by the enterprises' own funds. The non-appointed producers, however, were mainly dependent on their own funds.

The administrative allocation of funds led to the propensity of firms to import more capital goods. The programme of renovating existing enterprises with imported technology provided an opportunity for firms to get investment funds with subsidies. In addition, firms importing capital goods within the technology renovation plan enjoyed tariff exemptions.[5] Under this circumstance, the objective of firms in technology transfer projects was to obtain as large as possible a share from the total investment funds made available by the central planning organs.[6] Importing more capital goods which were technologically much better than domestic ones became the first choice of Chinese firms. Once firms possessed the capital goods, they could enjoy at least a share of the income generated. Furthermore, these assets were still the state's assets. The government had the obligation to help firms improve their utilization performance. Therefore, in the case that the imported capital goods had not been effectively utilized, firms could ask for additional funds to remove the bottlenecks. In addition, a firm with more fixed assets could apply for more working capital from state banks at a subsidized interest rate. In the investigation, there are three factories (NDWF, NWF and TJWF), the value of whose imported capital goods was much more than their total fixed assets before the technology import projects, due to the above mentioned reasons. However, these imported capital goods were not used effectively, because the three recipient firms did not have appropriate technological capabilities. This led to the unsatisfactory performance of imported technology and difficulties in future assimilation.

As a result of reforms, during the mid-1980s, bank loans began to replace government grants as the major financial resource for SOEs to carry out their investment projects (see Chapter 3). Furthermore, the CMRS required that

Table 5.15 Financial sources of assimilation projects

Number of firms whose assimilation funds are:	*Government grants*	*Bank loans*	*Firm's own*
100%	0	0	2
70 to 100%	0	6	2
30 to 70%	0	11	11
0 to 30%	0	1	5
0	20	2	0

Source: Own interviews.
Note: The number of firms which conducted additional import projects and/or R&D projects is 20.

SOEs had to use a specified part of their retained profits to finance their own technological activities. Therefore, as shown in Table 5.15, none of the sample firms obtained government grants for their assimilation projects. Nevertheless, all state owned and collective firms somehow managed to get the subsidized technological renovation bank loans for their assimilation projects. The only two firms which entirely relied on their own resources were TVEs. Moreover, if the assimilation projects of state firms were brought in line with the government economic plan, they were likely to be allocated more bank loans. As the result of the CMRS requirement, there were no firms which relied entirely on subsidized bank loans. From Table 5.15, it can be seen that for the majority of the sample firms (eleven out of twenty firms), the firms' own funds and bank loans each accounted for between 30 per cent and 70 per cent of their total investment.

In order to find the role of government policy in the early 1990s, two questions were included in the questionnaire to determine the importance of the factors affecting the technological decisions of the sample firms (Tables 5.5 and 5.6). The responses of firms show that the government is still a relatively important consideration in the assimilation decisions of firms, although its impact is decreasing. Firms complained that government did not provide sufficient preferential policy measures to help them carry out assimilation projects. In this respect, they mainly meant that the government did not give them sufficient subsidized financial resources which prevented them from investing more in assimilation projects. This, together with the high rating given to the availability of bank loans as a positive factor, reflects the chronic reliance of state firms on the government. However, the modest ratings of the government requirement on investment in technological activities among the positive factors and of the government restriction to control the types of investment projects among the negative factors (Tables 5.5 and 5.6) indicate that SOEs had gained more autonomy as the result of economic reforms. This shows that SOEs were less responsive to government technology policy than before, even though the policy was still carried out through administrative measures like it used to be.

The role of market competition

As was discussed earlier, the market for tape recorders in China changed dramatically from the mid-1980s as a result of more firms entering the industry and the liberalization of the product market (see Chapter 3). In the mid-1980s, the market for tape recorders was less competitive because the demand for tape recorders was far greater than the supply and the domestic market was protected by the government from foreign products. Furthermore, SOEs were in a privileged position because they were the first firms to gain access to imported technology and equipment. Consequently, the pressure from market competition to force SOEs to invest in cost reduction and new product development was weak, because they could transfer the high production costs to buyers. The less competitive market together with the subsidized financial resources was responsible for the propensity of the sample SOEs to adopt a capital intensive technology.

In the late 1980s, the competition in the market for tape recorders became increasingly fierce. At first, many SOEs supported by local governments entered the sector. Induced by the high profit margins of the past, they repeatedly imported capital goods for the production of tape recorders. Second, due to economic reforms, nonstate firms had developed rapidly since the mid-1980s and began to enter the tape recorder industry. Consequently, the market for tape recorders was over-supplied and the competition in this market forced the price to decrease substantially.

Nevertheless, the impact of market competition on the technological decisions of the sample firms was mixed. As shown in Table 5.5, pressures from competitors and customers were given relatively high ranks as positive factors to encourage them to invest in technological activities. However, competition from imported tape recorders and from the over-supply of domestically produced tape recorders (due to the duplicated imports of production facilities) were listed as important negative factors that discouraged the sample firms from investing in technological activities to assimilate imported technology (Table 5.6). This can be explained by the argument that SOEs were highly sensitive to the short term profits they could make from their investment. This argument can be inferred from the high rankings given to the two positive factors of the growth of market demand for their products and the high profit margin for their products (Table 5.6).

The above data show that earning profits became an important consideration in the technological decisions of SOEs. However, facing the over-supplied market, the sample firms performed poorly due to the incompetence of SOEs in competing with nonstate firms. Consequently, they were commonly short of funds to maintain and improve their technological capabilities. According to the sample firms, their major concern in making assimilation efforts was investment financing (Table 5.6). The lack of financial resources is listed as the primary negative factor. In the investigation, firms said that shortages of funds were the most serious problem they had.

Facing poor performance and reduced government subsidies, their major concern was earning incomes for their employees, rather than thinking about long term technological investment (see discussion earlier on the deterioration of technology capabilities of the sample firms). In addition, it was increasingly difficult for the sample firms to obtain bank loans to carry out technological projects, because after the late 1980s, the tape recorder market became a buyer's one and the profitability of tape recorder production was very low. Local governments and the banks were thus extremely reluctant to provide credits to tape recorder producers.

Therefore, increased market competitive pressure did not force the sample firms to invest in cost reduction and new products development. Rather, the incompetence of SOEs in market competition led to the poor economic performance which further reduced their financial ability to invest in technology.

The incompetence of the sample firms in market competition also reflects that their technological capability was not appropriate and thus could not pass the test of market competition. The technological capability of the sample firms was built in the past in a protected market with heavily subsidized government funds to fulfil the production task in a central planning system. Consequently, the sample firms tended to use capital intensive technology and to carry out as many production functions as possible by themselves (see case studies in Chapter 6). When the market competition increased, this type of technological capability became a disadvantage to the sample firms, because of their high production costs due to the lack of specialization and of the slow adjustment to changes in market demand due to the rigid specification of assets. Therefore, market competition not only affects the technological behaviour of firms, but also acts as a selection mechanism for appropriate types of technological capabilities.

CONCLUDING REMARKS

This general survey of technology import projects and their performance of the sample firms gives us a broad picture of the process of technology transfer by state owned tape recorder enterprises. It reveals that the performance of technology import projects by Chinese state enterprises was not satisfactory in terms of either the creation and enhancement of technological capability or the economic outcomes of the sample firms as a result of technology import projects.

Data about the fulfilment of the planned targets of technology import projects show that the sample firms as a whole managed to produce the planned output in the mid-1980s when most of the sample enterprises had just completed their technology import projects and the market for tape recorders was dominated by sellers. However, they did not achieve their profits target, because of their high production costs. As competition in the market for tape recorders increased from the late 1980s, a large proportion of the sample firms were forced to reduce their output. Their profits dropped even more dramatically. This meant that faced with competition from non-state enterprises, the economic outcome of state enterprises in utilizing imported technology deteriorated.

As far as technological capability improvement is concerned, the sample firms also performed unsatisfactorily. Generally speaking, the sample firms were passive users of the imported technology. Although a large part of the sample firms reported that they had installed the imported technology on schedule and utilized it at full capacity, a few firms had difficulties in the implementation of imported technology, due to problems in the technology acquisition period. Nevertheless, according to the questionnaire interview information, most of the sample firms had not modified or improved the imported technology significantly, although they tried to localize the production of most parts. Moreover, only a few firms reported that they had developed new products independently and had made their own innovations based on the imported technology. This shows that the sample firms mostly remained at the 'lower level' of technology assimilation, if their technological capability is evaluated based on the sequence model of the analytical framework. That is to say that their technological capability as a result of the technology transfer projects was limited to producing tape recorders based on the imported product technology without significant modifications.

In the mid-1980s, when tape recorders were in a seller's market, being able to simply use the imported facilities allowed the sample firms to earn high profits (although not as high as planned). However, as more firms entered the market for tape recorders and the market became competitive after the late 1980s, SOEs in the tape recorder industry declined because they failed to cut costs and to create market niches based on technological innovation. The poor economic performance of the sample firms in turn reduced their financial ability to invest in the maintenance and improvement of their technological capabilities gained from the technology transfer in the past. This vicious circle led to a further deterioration in the technological capabilities of many of the sample firms, in the sense that a large proportion of imported facilities lay idle and many technicians left the sample firms to do unrelated work elsewhere.

The performance of the sample firms in transferring imported technology is closely related to their behaviour in technology acquisition, utilization and assimilation. In the mid-1980s, when most of the sample firms undertook their technology import projects, their technological behaviour could be characterized as capital goods oriented. They preferred assembly lines and manufacturing equipment to 'software' technologies of blueprints and technical know-how. This kind of technological behaviour was affected by the policy and institutional environment at that time. First of all, state enterprises were granted more autonomy in carrying out technology import projects and allowed to retain a proportion of the profits earned. Second, the costs of financial resources to state enterprises were relatively cheap, because they were allocated subsidized funds through the state plan of the technological renovation programme. Third, capital goods allowed state enterprises to have a quick return in the market for tape recorders, where they enjoyed a monopolistic position, because the tariff was high and state enterprises were the first to import the production facilities.

Due to this distorted incentive structure, state enterprises, together with local governments, tended to put their projects into the state plan and to pay less attention to the costs and the feasibility of the projects. Consequently, some firms had difficulties in implementing the imported facilities on schedule and some firms could not utilize the imported equipment at full capacity. Furthermore, the propensity of state firms to import capital goods rather than technological know-how led to the difficulties in assimilating the imported technology later on.

This technological behaviour of state firms in acquiring foreign technology deviated from the objective of the government policy on technology imports which aimed to improve the effectiveness of technology imports in lowering costs. This deviation was a manifestation of the agency problem in technological activities of state enterprises. This agency problem was important at the first stage of economic reform during the mid-1980s (see Chapter 3) for the following three reasons. First of all, the reform measures of 'profit sharing and enterprise autonomy enlargement' provided unprecedented incentives for state enterprises to pursue their own objectives of income maximization (Lee 1991; Walder 1989; Huang 1990; and Tam 1992). Second, while the governance of state enterprises was still carried out through the administrative hierarchy designed for central planning (Jefferson and Rawski 1994), the inherently ineffective control from the central government as the principal was further weakened. Third, the financial resources of technology import projects were still largely subsidized and allocated administratively. Under this institutional environment and operating in a monopolistic market, acquiring a capital goods intensive technology became a natural choice of state enterprises because this would help them to achieve their own objective of income maximization.

The interview information shows that after the wide implementation of the CMRS in the late 1980s when the sample firms proceeded into the assimilation phase of their technology transfer process, they made very limited long term investment in learning and improving the imported technology. If they had to invest in technological activities as required by the contract, their efforts were biased towards purchase of capital goods. This kind of technological behaviour shows that the sample state firms were short term orientated due to the institutional arrangements governing the operation of state enterprises. First of all, the CMRS contracts, which were normally for three to five years, actually gave state enterprises a kind of de facto property rights to the revenue generated from the industrial assets. However, the rights were neither exclusive nor transferable. Second, the clause in the CMRS contracts regarding technological investment by state firms was difficult for the government as the nominal owner of the industrial assets to enforce. Third, state enterprises were still subject to strict government controls, especially with respect to employment and financial management (Harrold and Lall 1993: 42). Finally, related to the rigid labour system, state enterprises not only could not respond to market signals by adjusting the size of their labour force, but also had to take the burden of maintaining full employment and providing various social welfare benefits for their workers.

Due to these problems, the sample state firms tended to invest less in their long term technology development, even though they faced an increasingly competitive market and 'a considerable reduction in the softness of budget constraints' (Jefferson and Rawski 1994: 32). Consequently, the sample state enterprises could not retain their technological advantages. Their poor economic performance further reduced the financial resources which could be used in technology development. As a result of being unable to compete in the market, many sample firms lost the technological capabilities established through the previous technology import projects.

The technological behaviour of the sample firms after the late 1980s showed the fundamental difficulties in reforming state enterprises within the framework of government ownership designed for a command economy. Specific to our case, the reform programme until then had not solved the problem of how the industrial assets of state firms should be used efficiently and improved effectively without resorting to bureaucratic power. Due to the nature of industrial assets, especially that of intangible industrial technological assets, it was very difficult to delineate clearly the property rights to these resources through the form of a responsibility contract. Enlarging enterprise autonomy while maintaining government ownership over the industrial assets through the CMRS system actually further blurred the property rights structure. Due to the problem of common property (see Chapter 2), the quasi-rents of technological capability of firms dissipated, as the participants to the contract tended to limit their investment in technology.

The general questionnaire interviews, however, only point out that the current institutional arrangement is an important factor affecting the technology transfer process of the sample firms during different periods of the economic reforms. They do not tell us how the changes in institutional environment affected the technological behaviour of state enterprises. The next chapter will, through detailed case studies, describe how technology import projects and assimilation activities were organized and implemented.

6 Technology import projects of four SOEs

From the thirty-one sample tape recorder firms, four firms have been selected for in-depth case studies. They are Canzhou Recorder Factory (CRF), Westlake Recorder Factory (Westlake), Weihai Recorder Factory (WRF) and Wuxian Wireless Factory (WWF). The first two firms are producers of tape recorders and the other two are manufacturers of tape decks (see Table 6.1). All these four enterprises are state owned and under the supervision of local government agencies. In common with other Chinese SOEs, they all implemented the reform measures of tax-for-profits during the early and mid-1980s and then subsequently adopted the responsibility system from 1987 to the early 1990s (see Chapter 3). The detailed case studies in the following two chapters are intended to reflect the complexity of the circumstantial context and to exemplify the impact of institutional arrangements on the process of technology transfer in SOEs.

Among the four firms, CRF and WWF were relatively larger in size and performed better in using imported technology, especially during the 1980s, than did WRF and Westlake (Table 6.1). As will be seen in the detailed case studies of individual firms, this is mainly because the first two firms had better management and technological capabilities. During the mid-1980s, SOEs were almost the only producers in the tape recorder industry. Therefore, their performance in transferring foreign technology was mainly affected by the different microeconomic and technical factors, such as the management and technological capabilities of individual state firms.

None the less, the case studies show that the technological behaviour of these four firms in importing and using foreign technology and equipment was similar, because they were governed by the same institutional arrangements. They preferred capital goods to licensed technical know-how when they imported technology, and they placed less emphasis on, and invested less resources in learning how to use this imported technology more effectively and efficiently. Consequently, when more and more non-state firms entered the market and SOEs lost many of their privileges after the late 1980s, the performance of all of the four sample firms declined, no matter how well they had performed in the mid-1980s. This reflects the importance of the macroeconomic factors of institutional arrangements

Table 6.1 Background of the four sample SOEs

	CRF	*Westlake*	*WRF*	*WWF*
Major products before TRP[a] production	Radios and wireless instruments	Radios and wireless instruments	Measuring and cutting tools	Wireless transceivers and radios
Year of starting TRP production	1979	1978	1979	1979
Year of importing TRP technology	1983	1983	1982	1982
Number of major technology import projects	2	3	2	3
Major TRP (1990–3)	Portable tape recorders, Hi-fis and walkmen	Portable tape recorders	Tape decks	Tape decks and tape recorders
Number of employees (1993)	2,987	460	929	2,542
Total turnover (1993) (m RMB ¥)	228.3	11.7	29.9	131.4[b]
Percentage of TR products in total turnover (1993)	97.3	71.2	54.1	100[b]

Source: Own investigation.
[a] TRP stands for tape recorder products and, in this table, includes tape recorders and tape decks.
[b] This figure includes the turnover of CD decks which WWF began to assemble from 1992.

and market competition in affecting the technology transfer process of individual firms.

CASE ONE: CANZHOU RECORDER FACTORY

The Canzhou Recorder Factory (CRF), located in Canzhou, Jiangsu Province, was established in 1978 through a merger of a radio factory and a wireless instrument factory by the Municipal Bureau of Electronic and Instrument Industry of Canzhou.[1] In the following ten years, the new organization conducted two major technology import projects and developed very quickly. By the late 1980s, CRF had become one of the major tape recorder producers in China.

The technology import projects of CRF in the 1980s

The development of CRF in the 1980s can be divided into three stages. The first stage was from 1979 to 1983, when CRF imported four assembly lines through the channel of a compensation trade agreement with a Japanese tape recorder firm, and assembled tape recorders from imported CKD kits. Under this arrangement, the Japanese firm provided the assembly lines and CKD kits of tape recorders to CRF and agreed to buy back the final products which were used to repay the costs of these assembly lines.[2] As a result of the compensation trade agreement with the foreign supplier, CRF gained experience in assembling tape recorders and established a production facility, of which the assembly capacity in 1983 was 300,000 sets of portable single deck tape cassette players. However, the technological capability of CRF established during this stage was very limited in the sense that the enterprise had not improved its technological capabilities in product design and plastic parts production, but had only gained the capability to assemble tape recorders. The firm's prior experience, though relevant, was limited to the production of portable semi-conductor radios, for which the electronic circuit was much simpler and the plastic parts less sophisticated than those of tape recorders.

The second stage was from 1983 to 1986. In 1983, encouraged by the new policy of renovating existing enterprises with imported technology, CRF proposed and carried out a technology renovation project aimed at the localization of tape recorder production with a total investment of RMB ¥8.79 million (Table 6.2). The purpose of the project was to establish an annual production capacity of 650,000 sets of tape recorders. The major contents of the technology import project included: (1) technological renovation of existing assembly lines; (2) importation of high precision mould-extruders; (3) erection of a coating and printing production line for plastic parts; and (4) importation of measuring and testing instruments.

During the second stage, the focus was concentrated on establishing a basic production capacity for tape recorders. Therefore, new product development was limited to copying foreign products. Moreover, mould-making was not on the agenda. The firm still relied on foreign suppliers for large and sophisticated moulds and only produced simple and small moulds by itself. Nevertheless, in 1986 CRF developed and produced several types of high grade tape recorders, such as those with functions of automatic programme selection, volume balancing and graphic equalizing. Moreover, the performance of the project seemed to be good. In 1986, CRF overfulfilled the planned targets of output, turnover and profits set in the feasibility study. That was a satisfactory achievement as compared to the technology import projects of other tape recorder producers (see Chapter 4 and Table 4.10). Labour productivity also increased steadily. As a result, the total realized profits of CRF increased by 50.3 per cent from 1984 to 1986.

The third stage started in 1986 when CRF proposed its second technology import project, which aimed at further enhancing its technological capability. As a result of the first technology renovation project, CRF expanded its assembly capacity and established a production facility for plastic parts. However, the production of the firm was unbalanced because, as compared to its acquired production capacity in parts production and final products

Table 6.2 The performance of the technology import projects of Canzhou Recorder Factory

	1983	*1984*	*1985*	*1986*	*1987*	*1988*	*1989*	*1990*	*1991*	*1992*	*1993*
Actual performance											
1 Turnover (RMB ¥m)	72.64	81.26	117.68	157.11	183.26	207.5	233.66	225.84	182.6	218.34	228.3
2 Total profits (RMB ¥m)	7.52	9.21	12.55	14.34	16.62	17.26	15.39	8.52	3.21	4.24	5.52
4 Net fixed assets (RMB ¥m)	—	24.87	31.9	42.9	50.2	58.6	60.95	61.2	62.8	62.5	61.3
6 Number of employees	—	2,398	2,455	2,667	2,989	3,035	3,067	3,180	3,175	3,159	2,987
Planned targets											
First project (total investment RMB ¥8.79 million)											
Planned turnover (RMB ¥m)			122	155	170						
Planned total profits (RMB ¥m)			12.5	14	16						
Second project (total investment RMB ¥9.87 million)											
Planned turnover (RMB ¥m)					180	205	235	250			
Planned total profits (RMB ¥m)					16.5	17.5	19.5	21.2			

Source: Own interviews.

assembly, the ability of the firm to develop new products was weak and the quality of its plastic parts was relatively poor in terms of the accuracy and decoration of its products. These problems prevented the firm from upgrading its products. Therefore, in the feasibility study for the second technology renovation project, CRF outlined two major tasks. The first task was the modernization of the firm's product design facility. The second task was to establish a mould-making facility and to renovate the coating and printing production line for plastic parts, so that the process of producing new types of plastic parts would be accelerated and the quality of plastic parts improved. This part of the second project was carried out by the division of plastic extrusion. According to the plan, these tasks would be undertaken during the period from 1986 to 1987 (Table 6.3). The total investment was RMB ¥9.865 million, which was covered by bank loans (75 per cent) and CRF's own funds (25 per cent). In addition, the provincial Economic Commission allocated a foreign exchange quota of US $2.11 million to the factory for the importation of equipment.

As a consequence of these technology renovation projects, CRF established a comprehensive and relatively modern technological capability that included new product development, plastic parts production and decoration, plastic mould designing and manufacturing, and audio and electronic inspection and testing facilities. During the period from 1985 to 1988, CRF launched a series of new products under the brand name of GLOBE, which became a famous trademark for domestically produced tape recorders. The firm was awarded the gold medal in a nationwide competition of audio

Table 6.3 The second technology import project of CRF

Total investment (RMB ¥m): 9.87				
Foreign exchange (US $m): 2.12				
A. Designing technology				
Total investment (RMB ¥m): 3.41				
Foreign exchange (US $m): 0.74				
B. Mould making and surface coating project				
Total investment (RMB ¥m): 6.46				
Foreign exchange (US $m): 1.39				
	1987	*1988*	*1989*	*1990*
Targets of mould making and parts decoration:				
Mould making capacity, mould output (set)	520	650	720	720
Plastic parts decoration, parts output (1,000 pieces)	6,420	6,500	6,800	6,800
Targets of the whole second project:				
Turnover (RMB ¥m)	180	205	235	250
Total realized profits (RMB ¥m):	16.5	17.5	19.5	21.2

Source: Own interviews.

products in 1987 and was named as a high quality enterprise by the State Economic Commission in 1988, because CRF's MTBF (minimum time before failure) and the acceptable rate of final products, which are the two major quality indicators of tape recorders, were among those of the top firms. Moreover, the turnover and the profits of CRF increased by 127.4 per cent and 87.4 per cent respectively between 1984 and 1988 (Table 6.2).

In the meantime, the fixed assets of CRF increased rapidly. As a result of the two technology import projects, CRF acquired a great deal of modern machinery and equipment. In addition, because of the enterprise reform, CRF adopted the tax-for-profits system during the period 1984–1987, which allowed the firm to retain a larger portion of the total realized profits than before. Thus, the firm had the financial ability to invest in fixed assets expansion, and its net fixed assets increased by 135 per cent from RMB ¥24.9 million in 1984 to RMB ¥58.6 million in 1988.

The CMRS system and the performance of CRF after the late 1980s

In early 1988, CRF adopted the Contract Management Responsibility System (CMRS) and signed a contract with its supervisor, the Bureau of the Electronic Industry (BEI) of Canzhou Municipality, which represented the government as the owner of state assets. The contract stipulated the targets which CRF had to fulfil for the period from 1988 to 1990, when the Seventh Five Year-Plan would end. Similar to other contracts under the CMRS, there were three major parts in the contract (the profit sharing scheme, the technological investment activities, and the wage and bonus scheme). The key component, however, was the profit sharing scheme, because the other two depended on it.

According to the contract, the total realized profits of CRF were to increase by 7 per cent annually for the three years from 1988 through 1990, as a result of completion of the second technology import project in 1988 (Table 6.4). Based on the forecast figure of profits agreed to by CRF and the BEI, CRF would hand over 55 per cent of its total realized profits to the government.

The increased profits and production capacity of CRF during the mid-1980s were mainly attributable to the privileged position of SOEs. Entering the late 1980s, however, the market for tape recorders became increasingly competitive, because more and more firms entered the field. In addition, as economic reforms proceeded, the privileged access of the sample SOEs to subsidized funds and foreign exchanges were removed. Under this situation, the inherent problems of SOEs in importing foreign technology, such as the bias towards capital goods and the lack of learning efforts, soon led to CRF losing its competitive advantage, because the technological capabilities of the firm established as the result of the two technology import projects focused only on production capacity. CRF did not possess a technological advantage over other firms, and thus did not enjoy a market niche based on high-grade

Table 6.4 CMRS in Canzhou Recorder Factory (¥m)

	Planned profits 2nd project	Actual profits	Planned profits of CMRS contract	Profit remittance		Profit retention	
				Target	Actual	Target	Actual
1985	—	13.55	—	—	n.a.	—	n.a.
1986	—	14.34	—	—	n.a.	—	n.a.
1987	16.5	16.62	—	—	9.1	—	7.48
1988	17.5	17.26	17.8	9.8	9.93	8	7.34
1989	19.5	15.39	19	10.45	9.05	8.55	6.34
1990	21.2	8.53	20.3	11.15	5.25	9.15	3.28
1991	—	3.22	—	5.25	1.2	—	2.02
1992	—	4.24	—	5.8	1.55	—	2.69
1993	—	5.52	—	6.4	2.65	—	2.87

Source: Own interviews.

products which other domestic firms could not make. Relatedly, due to the problem that CRF established an all-round production capacity and carried out most production tasks by itself, ranging from mould-making, parts production and furnishing, to final product assembling, it was difficult for CRF to change its product mix quickly to suit the shifts in market demand promptly. Furthermore, the fact that CRF carried out nearly every production task by itself led to high production costs, because a large part of the expensive imported capital goods was not utilized at full capacity.[3] In addition, the lack of specialization prevented CRF from enjoying the economies of scale that a capital intensive firm usually enjoys. Consequently, from 1989 the turnover of CRF stagnated and its profits declined sharply (Table 6.2).

Because CRF could not respond promptly to the shifts in the tape recorder market in the late 1980s, its actual performance was far less satisfactory in terms of the fulfilment of the contracted target of total realized profits. Although the government greatly reduced the profits remittance in 1990 when the total realized profits of CRF declined sharply, the retained profits of the firm decreased even more severely, because it still had to hand over a certain amount of profits to the government. Therefore, in the second run of the CMRS which started in 1991, the government allowed the targets of the total realized profits and profits remittance to be adjusted downwards greatly to help the firm to overcome its difficulties (Table 6.4).

The dramatic decline of profitability together with the increasingly serious short term behaviour of SOEs under the CMRS had a negative impact on the technological investment activities of CRF in the late 1980s. As was mentioned earlier, the fixed assets of CRF increased by RMB ¥33.7 million from 1984 to 1988. By contrast, however; the increase of fixed assets of CRF after 1989 was very slow. From 1989 to 1990, net fixed assets increased by only RMB ¥2.6 million, which was much less than the technological investment targets of RMB ¥6.64 million as stipulated in the first contract. Entering the 1990s, the situation became even worse. In 1992 and 1993, the net fixed assets of CRF decreased by RMB ¥0.3 million and RMB ¥1.3 million respectively. This means that the investment made by the firm during this period even did not compensate for the depreciation of its fixed assets. The lack of investment in technology prevented the firm from updating its production facilities and further prevented the firm from improving its performance.

CASE TWO: WESTLAKE RECORDER FACTORY

Westlake Recorder Factory (Westlake) is a medium-sized state enterprise located in Hangzhou, the capital of Zhejiang Province. The major products of the firm in the 1980s were low- and medium-grade tape recorders. However, in 1993, the firm greatly reduced its production of tape recorders because from 1990 onwards the firm had made losses for four consecutive years and was facing a very difficult situation.

Technology importation and utilization in the 1980s

The predecessor of Westlake was Hangzhou No. 4 Wireless Factory which was set up in 1965 to produce portable semi-conductor radios and other wireless instruments for industrial uses. In 1978, No. 4 Wireless set up a tape recorder branch and began to assemble tape recorders from imported CKD kits. However, because it used the existing production line for radios to assemble tape recorders, the scale of tape recorder production was small and the firm's performance was poor. In 1982, in order to promote the production of household electronic visual and audio products in the Hangzhou area, the municipal government set up the Hangzhou Broadcasting and Television Industry Corporation (HZBTIC). The tape recorder branch was separated from No. 4 Wireless as an independent factory with the name of Westlake Recorder Factory, and became a member factory of HZBTIC. Within HZBTIC, there was another tape recorder producer, Qiuying Wireless Factory (Qiuying) which was a small state owned factory previously under the supervision of the government of a district of Hangzhou Municipality.

With the technical and financial assistance of the Corporation, Westlake and Qiuying each imported a tape recorder assembly line and a complete set of production equipment and inspection instruments in 1983, through a Hong Kong consulting company. The major key pieces of equipment included a wave soldering machine and a central signal generator. In addition, Westlake procured a number of domestically manufactured plastic extruders and metal processing machine tools, so that it could expand its production capacity of plastic parts and establish a basic mould-making capability (Table 6.5). During the period 1983–4, under the coordination of HZBTIC, Westlake and Qiuying each assembled the two same types of desk tape recorders,[4] of which the key components were largely imported.

Table 6.5 Technology import projects of the Westlake Recorder Factory

Year	*First project 1983*	*Second project 1985*	*Third project 1987*
Major contents			
a. Imported equipment	One assembly line; wave soldering; central signal generator.	One assembly line; CKD kits (30,000 sets); inspection instruments.	One production line of surface coating.
b. Domestic equipment	Plastic extruders; metal processing machine tools.	—	—
Technological capability being gained	Assembling; plastic parts manufacturing; mould making.	Assembling; product design imitation.	Operational.
Comments	Unsuccessful due to poor project management and facility relocation.	Profitable, because of a high demand in the market; technological capability was limited to imitation of imported products; no cost reduction programme.	Unsuccessful due to deteriorating overall performance.

Source: Own interviews.

Although their products were profitable, the production of tape recorders in Westlake developed very slowly, because HZBTIC concentrated on the development of television receiver production. Consequently, Westlake only obtained a relatively small amount of financial resources from the government. In late 1984, in order to promote the local production of television receivers, the municipal government decided to establish the Hangzhou General Factory of Television (General TV) by separating the tape recorder production from HZBTIC. The two tape recorder producers of Westlake and Qiuying in HZBTIC were thus merged into an independent firm, which was still called Westlake Recorder Factory, under the supervision of the Bureau of Instrument-making Industry. Because the original premises of the former Westlake were to be used by General TV, Westlake relocated all its equipment and machinery to Qiuying, which had enough available space to accommodate the entire production facilities. As a result of the merger, the new Westlake became the largest audio goods producer in the Hangzhou area. It possessed two tape recorder assembly lines, one tape recorder/radio mixed assembly line, a plastic parts production plant and a set of mould making facilities.

However, the performance of Westlake after the merger was poor. First of all, the relocation of equipment to Qiuying and the re-arrangement of the equipment layout took a long time. Second, there was a conflict between management personnel from the two predecessor firms. Consequently, production planning and coordination did not proceed smoothly and worker discipline was loose. Third, the old products of desk tape recorders were outmoded and Westlake failed to launch new products, because the firm was unable to solve many technical problems in the manufacturing of the plastic parts, especially the outer case, of its planned new recorder.

These problems at Westlake reflected that the outcome of previous technology import projects was not satisfactory. Although it had been in operation for several years, Westlake had not established the technological capability to produce tape recorders independently. Under the HZBTIC, Westlake lacked decision making autonomy in formulating its own development strategy and in proposing technology import projects. Furthermore, the financial resources were allocated between the two predecessor tape recorder firms through administrative channels. This led to the situation that both firms imported similar assembly lines. As a result, although after the merger Westlake became the largest producer of tape recorders in the Hangzhou area, its technological capability did not go beyond the assembly of imported components, because neither the former Westlake nor Qiuying had attempted to generate technological capabilities in parts production and new product development.

Faced with this situation, the new management of Westlake determined to carry out another technology import project. In 1985, they managed to bring the project into line with the provincial technology import programme and obtained a technology renovation bank loan. However, what they actually imported from a Japanese company were 30,000 CKD kits of a stereo AM/FM dual-deck portable tape recorder and a number of key pieces of equipment and inspection instruments to assemble these CKD kits. Westlake made high profits from this project, because the quality of the imported CKD kits was high and the style of the imported product was new. More important, this was one of the last imports of tape recorder CKD parts by state firms for the purpose of domestic sales.[5] Therefore, Westlake could sell these tape recorders at a higher price.

Because this type of tape recorder was popular, Westlake developed and produced several types of tape recorders by copying the imported product in the following two years. Since the market for tape recorders at that time was a seller's market, the firm made a reasonable profit, although its production costs were not reduced as its cumulative output increased. In 1987, Westlake undertook an additional technology import project of plastic parts surface coating and printing. This project was financed by bank loans (65 per cent) and the firm's own retained profits (35 per cent) and included the importation and installation of a coating line and a number of printing machines.

It was hoped by Westlake that the coating and printing project would improve the appearance of its plastic parts and final products, and hence that the competitiveness of its tape recorders in the market would be increased.

However, because competition in the tape recorder market became increasingly fierce from 1988 (see Chapter 4) and because the quality of tape recorders produced by Westlake was not good and its brand name of Westlake was not well known nationwide, the firm could only sell its products in the local market. Due to the penetration of firms from other provinces into the local market, Westlake's market share was shrinking. As a result, the total realized profits of Westlake dropped significantly (Table 6.6).

In early 1988, Westlake signed a CMRS contract that covered the three years from 1988 to 1990. The contract stipulated that when Westlake overfulfilled the total profits target, it could retain as high as 75 per cent of the above target profits, as long as it gave priority to technology investment. Nevertheless, the contract did not specify what kind of technological activities Westlake should undertake in the contract period. In order to achieve the profits target and obtain more retained profits, Westlake expanded its output in 1988 and 1989.

In order to support the production expansion, it borrowed heavily from the bank to finance working capital. However, because the firm did not make sufficient effort to develop new products to suit the changing conditions of the tape recorder market, its sales did not increase as quickly as its output, and the total realized profits of Westlake decreased from nearly RMB ¥2 million in 1987 to less than RMB ¥0.4 million in 1989 (Table 6.6).

Because of the piling up of unmarketable final products and the deterioration of profitability, Westlake encountered a serious cash flow shortage and the bank refused to continue to lend working capital to it. Westlake was forced to cut its production of tape recorders significantly and incurred a loss of RMB ¥2.35 million in 1990. Although the government provided financial assistance, the firm still could not improve its performance. From 1990 to 1993, the firm incurred losses for four consecutive years and its cumulative losses during this period were RMB ¥12.3 million, which was 60 per cent more than the sum of Westlake's profits from 1985 to 1989. Consequently, after 1992, the total debts of the firm exceeded its total assets.

Table 6.6 Performance of Westlake Recorder Factory (RMB ¥m)

	Output value	*Total profits*
1985	15.89	1.68
1986	19.24	2.27
1987	21.35	1.96
1988	26.69	1.25
1989	29.22	0.38
1990	16.35	−2.35
1991	12.32	−4.21
1992	8.32	−3.25
1993	9.26	−2.21

Source: Own interviews.

Institutional restrictions on Westlake in extricating itself from operational difficulties

Faced with this deteriorating performance, the management of Westlake and its supervisory governmental agencies tried several approaches to 'vitalize the firm'. However, as we will see later, these methods were not effective, because they could not provide a resolution to the problems caused by the various constraints faced by Westlake as a state firm. Consequently, the firm could only maintain the *status quo*. Under these circumstances, not only could its imported technology not be assimilated and utilized effectively, but also its existing technological capability, such as that embodied in equipment, in people and in the organizational form of enterprise, was being lost. The reason is that when that firm always made a loss, the resultant shortage of financial resources meant that the firm was unable to renovate its capital goods and maintain its human capital. In this case, if the firm's productive assets could not be re-allocated to other uses in an efficient way, the situation became even worse and the technological capability of the firm soon lost its value, because industrial assets as well as technological capability embodied in human beings and organizations had a finite usable life.

One of the approaches that Westlake and its supervisory governmental agencies tried several times in order to rescue the firm was to form a joint venture, or to use some parts of the firm to form joint ventures with foreign companies. However, it was difficult for Westlake to find a cooperative partner, because Westlake's main operation was assembly and it lacked comparative advantages, such as a skilled labour force or market access. Compared with firms in Southern Guangdong Province, Westlake did not enjoy the advantages of short geographic and business distances which were important factors to attract foreign firms (mainly those firms from Hong Kong) trying to relocate their assembly activities. More importantly, foreign firms did not want to keep so many workers, because their purpose in relocating their production from abroad to inland China was to exploit cheap labour resources. However, Westlake and the local government had no ability to make an acceptable arrangement for those superfluous workers.

In 1992, the government tried to merge Westlake with Hangzhou General TV from which Westlake had been separated in the mid-1980s. General TV was interested in the premises and the assembly lines of Westlake which could be renovated to produce television sets. However, it did not want to take the responsibility for the debts of Westlake. Moreover, General TV was only willing to accept the young and skilled workers who accounted for less than 50 per cent of Westlake's 700 employees. Although the government put pressure on General TV, it still did not want to accept all of Westlake's employees, because General TV had an over-staffing problem of its own and an increased number of employees would have meant that the income of General TV's existing workers would be reduced.

One option was for Westlake to go bankrupt. To help the firm, the government exempted it from paying RMB ¥2.8 million of technological renovation bank loans. However, there was still RMB ¥9.5 million of losses on the books at the end of 1993. The firm owed its suppliers and the bank RMB ¥13.4 million, while its inventory of unsaleable tape recorders and intermediates were worth RMB ¥6.2 million and the book value of its fixed assets was only RMB ¥2.6 million. Consequently, the total debts of the firm exceeded its total assets by RMB ¥4.6 million. Furthermore, because Westlake had not undertaken any investment projects since 1988, it was unlikely that the firm could improve its performance in such a difficult situation.

Nevertheless, Westlake still had valuable assets which could be re-allocated to other uses. The market value of its buildings was estimated as high as around RMB ¥13 million, because of their large space and the downtown location. The imported equipment such as the coating and printing machinery and inspection instruments were advanced. Moreover, workers and technical personnel of Westlake had experience in assembling electronic goods. All these assets could be utilized to generate wealth, if they could be re-allocated efficiently. However, an alternative to bankruptcy was implemented, mainly for three reasons. First, due to ideological and bureaucratic reasons, the supervisory agencies opposed the option of bankruptcy. Second, the problem of how to make arrangements for workers, especially for those unskilled and retired workers,[6] could not be solved. Third, the problem of how to repay the debts could not be settled.

As a result, Westlake maintained the *status quo* after 1991. In 1992, in order to reduce losses, the firm asked many workers to retire before their official retirement age. Moreover, it allowed workers to go home to do whatever they could to earn incomes, if they were willing to receive only the basic living allowance from the firm. Facing the hopeless situation and the decrease in income, many employees, especially skilled and young workers and experienced management personnel, left the firm. As a result, the total number of employees reduced by 30 per cent from more than 700 people in 1990 to less than 500 in 1993.

Because of the reduction of orders, Westlake was forced to close two of its four assembly lines. Of the remaining two assembly lines, one was used to assemble tape recorders and another one to assemble radios. In order to cut costs and earn incomes, Westlake divided the firm into several small independent accounting units and asked them to earn their own incomes. Furthermore, it rented out its buildings as warehouses and leased a part of its unused machinery to TVEs to earn incomes. Nevertheless, there were still many pieces of imported equipment that were idle because the government did not allow the firm to sell its excess equipment.

CASE THREE: WEIHAI RECORDER FACTORY

Weihai Recorder Factory (WRF) is a state owned enterprise located in Shandong Province of North China. It is a member enterprise of the Weihai

Electronic Audio Industrial Company, and under the supervision of both the Electronic Industry Bureau of Shandong Province and the Economic Commission of Weihai Municipality. In 1993, the factory had a total of 929 employees, among whom 8.8 per cent were technical personnel. The two major groups of products WRF produced were measuring tools and tape decks for cassette tape recorders. The turnover of tape decks was RMB ¥16.2 million in 1993, accounting for 54.1 per cent of its total sales.

The technology import project of WRF

WRF was established in 1979 by transforming a measuring tool plant to start assembling tape recorders and tape decks which, at that time, were new to the Chinese market and regarded as lucrative products with a large market potential. At the beginning, WRF assembled tape decks with imported CKD kits. In 1979, under an arrangement made by the General Bureau of Communication and Broadcasting Industry (GBCBI) of the Ministry of Machine Building and Electronic Industry[7] (MoMBEI), it signed a contract with a Japanese company which agreed to provide an assembly line and the related assembly technology to WRF. In return, WRF would import parts and components of tape decks from this company and assemble tape decks which then would be sold to domestic tape recorder factories. After three years production, WRF had established an annual assembly capacity of 500,000 sets and gained a certain amount of experience in assembling tape decks. Its actual output of tape decks in 1982 was 200,000 sets.

During the period from 1979 to 1982, the domestic production of tape recorders and tape decks in China was mainly undertaken through assembling the final products with imported parts and components. The import price of an assembled tape deck was 7 to 8 US dollars and the price of CKD kits was 4 to 5 US dollars at that time. To save badly needed foreign exchange, assembling tape decks with imported CKD kits through import substitution at the lowest end of tape deck production chain was the policy adopted by GBCBI,[8] which was in charge of coordination of tape recorder production throughout the country. Because individual enterprises then had no right to do business with foreign firms by themselves, all imports of equipment, technology and CKD kits were controlled by GBCBI. By organizing imports of CKD kits of tape deck and allocating the imported parts to domestic assemblers at a marked up price, GBCBI accumulated a large amount of foreign exchange.

GBCBI wanted for a long time to carry out technology import projects in domestic tape deck firms, in order to localize the production of parts and components for tape decks so that the huge amount of foreign exchange spent on importing two to three million sets of tape deck CKD kits every year could be saved. In 1982, the tape deck localization projects could be carried out for several reasons. GBCBI had the necessary funds, domestic enterprises had gained assembly experience and the central government had changed the technology import policy which now encouraged technological renovation of existing enterprises with imported technology (see Chapter 3).

WRF was selected by GBCBI to carry out a technology import project to produce the NTP series of tape decks. The NTP tape deck was developed by the Japanese company with which WRF cooperated for several years. The main characteristic of this tape deck is that its major components are a combination of plastic and metal parts, so that the total number of its parts is fewer and the assembly is simpler compared with those tape decks whose parts are almost totally metal.[9] However, the process technology of the metal/plastic combined parts of NTP tape decks is relatively sophisticated, because plastic parts are difficult to bond durably and solidly with metal parts.

With the support of GBCBI, WRF negotiated and concluded a technology agreement with the Japanese company in May 1982 for a three-year term from 1983 to 1985. According to the agreement, the Japanese company would sell the product technology of NTP tape decks to WRF and help it to localize the production of tape deck parts for one million sets annually. The planned localization rate was 30 per cent in 1983, 70 per cent in 1984 and 100 per cent in 1985. In addition, the Japanese company would provide technical advice to help WRF select and import the production equipment and tools (see Table 6.7).

The utilization of imported technology in WRF

The actual installation of imported equipment started in October 1983, which was half a year later than the planned starting time. Due to the lack of experience and poor management, the completion of the installation was delayed to April 1985, ten months after the planned completion time. According to the plan, WRF should have been able to increase its tape deck output from 300,000 sets in 1983 to one million sets in 1985 with a 100 per cent localization rate. However, the actual output was much less than the planned target (Table 6.7). Moreover, the localization rate[10] of parts was only 64.4 per cent in 1985.

There were four reasons for the unsatisfactory performance of the NTP project. First, WRF did not have the technological capability to absorb the product technology of the NTP tape deck selected by GBCBI. As a result, the rejection rate of some parts, such as the bracket for the driving wheel, was as high as 79 per cent. Consequently, the whole production process never stabilized, although WRF had already accumulated some experience in assembling this type of tape deck. Second, the project was completed nearly one year behind the planned schedule because WRF had to obtain approval from various governmental organs. Moreover, different departments gave different instructions that sometimes conflicted with each other. Third, WRF lacked adequate technological capabilities for both planning and implementing large investment projects. Thus, the production capacities of imported machine tools were not balanced and the whole installation was delayed. Finally, the

Table 6.7 The NTP tape deck project of the Weihai Recorder Factory

Major content of project:	Imports of NTP tape deck production technology and key pieces of equipment and instruments			
Planned installation schedule:	April 1983 to July 1984			
Actual installation schedule:	October 1983 to April 1985			
	1983	*1984*	*1985*	*1986*
Planned output (1,000 units)	300	600	1,000	1,000
Actual output (1,000 units)	312	556	623	675
Total planned budget (RMB ¥m)	9.378	100.00%		
Of the planned total budget:				
Foreign exchange (US $m)	1.849			
(Equivalent to RMB ¥m)	6.749	72.00%		
Domestic procurement and construction	2.629	28.00%		
Actual total investment (RMB ¥m)	11.02			
Of which:				
Foreign exchange (US $m)	2.104 (equivalent to RMB ¥7.80 million)			
Licensed technology	0.57	27.09%		
Equipment and instrument	1.349	64.11%		
Others (training and inputs)	0.185	8.79%		
Domestic procurement and construction (RMB ¥m)	3.22			
Equipment	2.17	67.40%		
Construction	1.05	32.60%		

Source: Own interviews.

management of production at WRF was poor. The management personnel did not know how to organize production on a large scale because WRF's past operations were in small batches. Consequently, the quality of parts was poor, leading to low quality of the assembled tape decks and high unit costs. Overstocks and shortages in the inventory of raw materials and semi-finished parts occurred simultaneously, which further hindered smooth operations.

Because the product technology of plastic/metal tape decks was not appropriate to WRF, GBCBI agreed in 1986 that WRF should change its product from plastic/metal tape decks to metal decks and allocated foreign exchange quota to the firm to help it undertake an additional technological renovation project. Because of the support of GBCBI, the project was placed in the provincial plan and allocated technological renovation bank loans (Table 6.8). The major purpose of the secondary technology renovation project was to import extra capital goods to enlarge and balance its parts processing capacity, especially the production capacity of metal parts. The imported equipment, which accounted for 93.5 per cent of the total foreign exchange spending, included high speed auto punches, a hot-cabin casting press, and die-making facilities. The planned output target of the project was

Table 6.8 The secondary tape deck technological renovation project of WRF

Major content of project:	Imports of extra equipment to balance the production capacity		
Installation time:	June 1986 to September 1987		
	1987	*1988*	*1990*
Planned output (1,000 sets)	1,000	1,200	1,200
Actual output (1,000 sets)	895	1,120	991
Actual total investment (RMB ¥m)	6.254		
Among total investment:			
Foreign exchange (US $m)	1.458 (equivalent to RMB ¥5.54m)		
Equipment and instrument	1.363	93.50%	
Others (training and inputs)	0.095	6.50%	
Domestic procurement and construction (RMB ¥m)	0.714		
Equipment	0.324	45.40%	
Construction	0.39	54.60%	

Source: Own interviews.

one million sets of tape decks in 1987 and 1.2 million sets in 1988 and thereafter. The motivation for both WRF and the local government, which was the major provider of financial resources for the secondary technology renovation project, was the high profit margin on tape decks. Because WRF already possessed some production capacity, even though it did not function well, it was hoped by the management that adding some other pieces of equipment to supplement the existing capacity might improve total performance and generate better profits.

The implementation of the secondary technology import project took fifteen months to complete. In 1988, WRF's output of tape decks reached 1.12 million units, which was 93.3 per cent of the planned target. If measured in terms of parts production capacity, WRF had finally reached the planned target of the original technology import project. However, as compared to the original plan, WRF spent three extra years and an additional investment of US $1.5 million to reach this goal. Moreover, due to the delay in the technology import project, WRF missed out on the best time to charge a high price and to seize a large market share during the period from 1984 to 1987 when the demand for tape decks was much higher than the supply. As is shown in Table 6.9, WRF did not fulfil its GVIO and the profits and taxes (P&T) targets after 1984. In 1987, its GVIO and P&T were 91 per cent and 69 per cent respectively of the revised targets. In 1988, although its output and GVIO increased by 25 per cent and 13.7 per cent respectively as compared with those in 1987, the P&T was reduced by RMB ¥0.1 million, reflecting increased competition in the tape deck market.

Since 1988, due to the increasingly fierce market competition, the performance of WRF has declined sharply (Table 6.9). The poor performance

Table 6.9 Performance of the NTP project and the secondary project of WRF

	1983	*1984*	*1985*	*1986*	*1987*	*1988*	*1990*	*1993*
Tape deck output (1,000 units)								
Planned[a] (NTP)	300	600	1,000	1,000	—	—	—	—
Planned (Secondary)	—	—	—	—	1,000	1,200	1,200	n.a.
Actual	312	556	623	675	895	1,120	991	604
GVIO (RMB ¥m)								
Planned (NTP)	12	24	40	40	—	—	—	—
Planned (Secondary)	—	—	—	—	39.2	45.8	44.2	n.a.
Actual	12.42	22.18	25.14	27.15	35.66	40.55	33.68	16.2
Profits and Taxes (RMB ¥m)								
Planned (NTP)	0.95	1.455	2.875	3.132	—	—	—	—
Planned (Secondary)	—	—	—	—	2.95	3.45	3.52	n.a.
Actual	1.06	1.52	1.64	1.86	2.04	1.95	0.98	–1.54

Source: Own interviews.

[a] The planned localization rate was 30% in 1983, 70% in 1984, and 100% in 1985 and afterwards.

Note: The figures of GVIO (Gross Value of Industrial Output), and profits and taxes only cover the tape deck business of WRF. The original plan did not give the targets of output, GVIO and profits and taxes for 1993.

negatively influenced the technological efforts of WRF. This is because, according to the CMRS contract signed in 1988, WRF was required to finance its technology projects out of its own production development funds, which came from its retained profits. Due to reduced profitability in the two years after the signing of the contract, the retained profits of WRF decreased even faster than the realized profits, because WRF had to turn over to the government a fixed amount of profits which was determined according to the average realized profits during the period from 1985 to 1987. Consequently, after 1988 WRF did not undertake large technological renovation projects, although there were some small R&D projects to modify its products.

The deterioration of technological capability in WRF

Entering the 1990s, the performance of WRF in utilizing imported technology went from bad to worse. The output of tape decks decreased by nearly half from 1988 to 1993. Although WRF tried to resume its production of measuring tools, its total GVIO was only a little more than 60 per cent of that in 1988. After 1991, WRF incurred losses. Under this difficult situation, WRF undertook some reform measures by splitting the enterprise into four divisions: measuring tools, metal processing, tape deck assembling and die-making. Each division was supposed to operate as an independent cost centre. They were allowed to do outside contracted work to earn their income, as well as to select their own suppliers to cut the costs.

However, this reform did not improve the performance of the whole enterprise. High costs, low quality and frequent delays in delivery prevented these divisions from winning outside contracts. In addition, the die-making division and the metal processing division, in particular, lost their previous work to provide intermediate inputs for other divisions in WRF, because the latter now switched to outside sources. For example, the metal processing division preferred to use dies made in Hong Kong, even though the cost of Hong Kong dies was a little bit higher than those produced by the die-making division of WRF, because the lead time was much shorter and the quality was higher for Hong Kong dies. Similarly, the tape deck assembling division preferred outside sourced CKD kits to parts made by the metal processing division. This caused a further reduction of the utilization rate of imported equipment and further eroded the technological capability of the firm. Because under this arrangement each division only considered their own short term interests, the technological development strategy for the whole enterprise was neglected and the technological linkage between divisions was weakened.

This new arrangement also failed to provide incentives to individual divisions as had been expected. Facing this difficult situation, divisions stressed the factors which they could not control. Moreover, they could not reduce the number of their employees. Therefore, WRF had to provide subsidies to the loss-making divisions (mainly the die-making division and the metal processing division) to help them to pay the wage bills and to provide fringe benefits for their workers, because for political reasons the incomes of employees of the loss-making divisions could not be too much lower than in other divisions. In order to reverse the trend of losing money, WRF planned to lease some workshops and facilities out to a township enterprise. However, the TVE intended to keep only one third of the staff members of these workshops. This meant that WRF would lose part of its technological capability, while still needing to increase its labour productivity, because it had to provide jobs for the workers who were not chosen by the TVE.

CASE FOUR: WUXIAN WIRELESS FACTORY

Wuxian Wireless Factory (WWF) is a medium-sized state owned factory. It is located in Wuxian, a prosperous city in Southern Jiangsu Province which is one of the fast growing areas in China. WWF was established in 1955 by merging several small privately owned radio assembling and repairing workshops. In the following twenty years, it conducted several investment projects financed by government grants. Throughout the 1950s and 1960s, WWF mainly produced wireless transceivers for military use. In the early 1970s, WWF started to produce portable semiconductor radios which soon became its major product and accounted for a half of its total sales.

In 1979, with the support of GBCBI (General Bureau of Communication and Broadcasting Industry, the Ministry of Machine Building and

Electronics Industry) and the Economic Commission of Jiangsu Province, WWF started to assemble tape decks with CKD kits imported from Japan. During the first years of the 1980s, WWF conducted three major technological renovation projects with imported technology and equipment to localize the production of tape decks. As a result, WWF became one of the major tape deck producers in China during the second half of the 1980s.

The three technology import projects

The first technology import project of WWF was in 1982 when GBCBI wanted to speed up the localization of tape deck production because the central government was going to increase the tariff on imported tape decks by the end of that year. The project was included in the 560 major technology import projects in the machine building and electronics industry during the period of the Sixth Five-Year Plan (1981–5). The target of the first project was to establish an annual production capacity of one million sets of tape decks within three years (Table 6.10).

The tape decks assembled by WWF during the early 1980s were metal ones, based on CKD kits supplied by a major Japanese tape deck manufacturer. In 1982, WWF had four assembly lines which were built with the assistance of the Japanese company, and the total assembly capacity was 350,000 sets annually, which could have easily been increased to one million sets by small improvements and converting one shift per day to two shifts.

Table 6.10 The production of tape decks of WWF

	Actual output (,000)		*Planned output (,000)*		
	Total	*Among total: CKD*	*1st project*	*2nd project*	*3rd project*
1979	25	25	—	—	—
1980	35	35	—	—	—
1981	100	89	—	—	—
1982	200	150	50	—	—
1983	680	380	200	—	—
1984	1,203	500	500	700	—
1985	1,860	800	1,000	2,000[a]	2,000
1986	2,300	—	—	2,300[b]	2,500
1987	2,800	—	—	2,500	3,500
1988	3,500	—	—	3,500	4,800
1989	4,200	—	—	4,200	5,500
1990	5,200	—	—	5,000	5,500
1991	3,500	—	—	—	—
1992	3,800	—	—	—	—
1993	3,750	—	—	—	—

Source: Own interviews.
[a] Key components need to be imported.
[b] All homemade hereafter.

However, WWF's capability to produce tape deck parts was very poor in terms of parts quality and production capacity. Therefore, it determined that the major task of the technology import project was to establish a comprehensive parts production capacity and a die-making capability. These were also the major aims of the second and the third technology import projects.

WWF's second technology import project was in 1984, when GBCBI asked WWF to further expand its production capacity of tape decks because the State Economic Commission and the Ministry of Machine Building and Electronics Industry were planning to stop tape deck imports in 1985. In response, WWF proposed and carried out another technology import project to enlarge the annual output target established in the first project from one million sets of tape decks to five million sets by the end of the 1980s. The total investment of the second project was RMB ¥8.54 million. In addition, the government allocated a foreign exchange quota of US $1.52 million for the factory to import equipment.

In 1986, GBCBI allocated a foreign exchange quota of US $1.8 million to WWF and helped it obtain a bank loan of RMB ¥6.5 million to launch the third technology import project which aimed to speed up the localization of tape deck parts production. The objective of the third technology import project was to fulfil the five million sets output target of the second project two years ahead of the original schedule of the second import project. Therefore, these three technology import projects can be seen as three consecutive stages of a single large project.

The performance of technology import projects in WWF

As a result of these three technology import projects, WWF became one of the most important tape deck producers in China during the second half of the 1980s. It possessed sixteen assembly lines with an annual assembly capacity of 5.5 million sets. Equipped with automatic, high speed and multi-position machine tools, punches, plastic mould-extruders and other types of specialized equipment, WWF was able to produce 95 per cent of the plastic and metal parts it needed. The factory also established a die/mould making centre which could design and manufacture all the dies and moulds it needed. Meanwhile, the quality of tape decks produced by WWF was good. The four major quality indicators of the tape speed, the shaking rate, the moment of force and the electricity consumption of WWF's products were similar to the quality of Japanese tape decks.

Although WWF had established an annual production capacity of 5.5 million tape decks and gained widespread technological capability – ranging from product updating, parts processing, die/mould designing and manufacturing, and machinery maintenance – the firm's performance was not satisfactory in terms of the fulfilment of the planned targets of output and profits (Table 6.11). Before 1988, WWF was far behind its planned increase in

Table 6.11 The performance of WWF's technology import projects[a]

	Output (,000 sets)		*Total sales (,000 Yuan)*		*Profits (,000 Yuan)*		
	Planned	*Actual*	*Planned*	*Actual*	*Planned*	*Actual*	*Actual profitability*[b]
1986	2,500	2,300	103,750	99,659	14,750	15,525	32.4
1987	3,500	2,800	145,250	117,964	20,650	16,464	29.0
1988	4,800	3,500	194,400	142,800	27,840	18,235	24.1
1989	5,500	4,200	209,000	162,078	30,800	14,574	19.7
1990	5,500	5,200	243,750	189,540	30,800	7,384	8.4
1991	—	3,500	—	123,935	—	−1,925	−2.7
1992	—	3,800	—	134,254	—	912	1.3
1993	—	3,750	—	131,438	—	2,100	3.0

Source: Own interviews.
[a] The planned figures were the targets of the third project.
[b] The profitability is the ratio of actual profits to the total capital employed.

output. After 1989, due to the slump in the market for tape recorders, the performance of WWF deteriorated dramatically. In particular, the profitability in terms of total realized profits to total assets of WWF declined very fast from more than 30 per cent in 1986 to only 8.4 per cent in 1990. In 1991, it incurred a loss of RMB ¥1.9 million. Although it adopted several measures to improve its performance, the profitability of WWF in 1993 was only 3.0 per cent.

The technological activities of WWF since the third technology import project

In late 1987, WWF began to adopt the Contract Management Responsibility System (CMRS). According to the contract WWF signed with its supervisory governmental agencies, WWF had the obligation to achieve a specified target of technological progress. Because it was very difficult to define improvement in technological capabilities, the technological progress target was stipulated in terms of the technological investment that the plant should carry out with its own retained profits.

In 1988 and 1989, after turning over the remitted profits which were stipulated on the basis of its actual realized profits in 1987, WWF invested its own retained profits as specified in the contract in several projects. These projects focused on adding some pieces of equipment and instruments to achieve the output target of the third technology import project.

In 1990, WWF roughly reached the output target for its third project. Output increased from 4.2 million sets in 1989 to 5.2 million sets. However, the realized profits of WWF, instead of increasing, decreased by nearly 50 per cent as compared with the previous year. The reason was that WWF failed to reduce its unit costs quickly enough to keep pace with the decreasing price of tape decks. In addition, the major product of WWF, that is tape decks for portable tape recorders, was no longer popular in the market and the firm was slow in updating its products to suit changing market demand.

That the output of WWF was unmarketable led to the problems of a heavy interest burden and lack of working capital for the firm. Those problems in turn forced the factory to reduce its production. As a result, it incurred a loss in 1991. Although the supervisory governmental agencies tried to help WWF by reducing its profit remittance quota in the second run of the CMRS starting in 1991, the factory's performance recovered only very slowly. Since 1990, WWF has not undertaken large technological investment projects, although there have been some small projects to develop new types of tape decks.

In order to extricate itself from its difficulties, WWF tried to diversify into the new generation of audio products, that is CD decks, to take advantage of its modern equipment. To enter the business, in 1992 WWF started a new technology renovation project to update its assembly lines by adding some key pieces of equipment. It was hoped by the firm that through assembling CKD kits of CD decks at first, it could gain experience in CD deck production and earn a high profit which would allow it to expand its CD production capacity. However, although WWF possessed advanced production facilities and had more technicians as the result of previous technology import projects, it could not compete in the CD player market with TVEs and FDI firms. This is because WWF no longer enjoyed privileges provided by the government as it did in the mid-1980s. On the contrary, TVEs and FDI firms located in special economic zones had the competitive advantages because they were close to Hong Kong and enjoyed preferential treatment from the government. As a result, the CD deck project was not profitable and the firm failed to obtain bank loans to carry out the second stage of its CD deck project to establish parts production capacity.

7 Impacts of macroeconomic factors on the technology transfer process in the case firms

THE PERFORMANCE OF THE FOUR SOEs IN TRANSFERRING FOREIGN TECHNOLOGY

The case studies of the four firms show that their performance changed as the economic reforms proceeded and the market for tape recorders became more competitive. The performance of the four sample firms in the importation and utilization of foreign technology was different during the mid-1980s. Two of them (CRF and WWF) performed better than the other two. Nevertheless, in the late 1980s, all four of the sample firms began to suffer from declining profitability. Faced with declining profits, they dramatically reduced investment in technology, resulting in further deterioration of their technological capabilities.

The performance in the mid-1980s

Of the four case study firms, CRF and WWF performed much better during the mid-1980s than Westlake and WRF in terms of both the establishment of production capacity and profitability. Consequently, CRF and WWF obtained more allocated bank loans from the government and developed faster than the other two. With the financial support of the government, they concentrated on establishing modern production facilities. Both of them identified several weaknesses in their technological capabilities. They thought these weaknesses could be overcome by importing more advanced capital goods. Therefore, after putting imported assembly lines into production, they conducted more technology import projects to establish their own parts production capacity by importing a large number of advanced machine tools. In addition, they built up modern product design and die/mould making facilities with imported advanced capital goods. They also stressed quality standards and technical advancedness in their products. To achieve a higher technical standard, they imported equipment and high precision inspection instruments.

As a result of a series of technology import projects, both CRF and WWF became major producers in the tape recorder industry with the most advanced machinery. They achieved a large part of their planned targets for project implementation and output increases. Furthermore, they established a comprehensive production capacity, ranging from die/mould making, designing, and parts manufacturing to final products assembling. During the second half of the 1980s, both of them became major producers in their sector. If the technological capability enhancement is assessed against the analytical framework developed in Chapter 2, these two firms achieved the 'higher level' of technology assimilation along the process of technology transfer.

By contrast, both Westlake and WRF performed poorly in importing and using foreign technology. Consequently, their economic outcomes were not satisfactory, even in a protected market. Both of them remained at the 'lower level' of technology assimilation. In the case of Westlake, the production activities of the firm were basically limited to assembling imported CKD kits and imitating the imported products. It did not invest much in assimilating the imported technology, and its technological capabilities in parts production and new products design were weak. Furthermore, the firm's technological capabilities to assemble outside sourced CKD kits and imitate imported products did not improve very much. Therefore, the efficiency with which the firm used the imported technology was low. Because it did not make efforts to modify both the imported process technology and product technology to adapt them to local conditions, Westlake failed to establish the technological capability to produce tape recorders independently and efficiently, let alone the technological capability to improve the imported technology and innovate based on the assimilated foreign technology (Table 6.5).

Similarly, WRF was not successful in implementing its technology import projects. It failed to master the product technology for plastic/metal decks, although a large amount of equipment was imported. After an additional investment financed by the government for the second technology import project, WRF managed to put the imported technology into production. Nevertheless, it was simply using the imported equipment and did not improve on the imported technology. Moreover, the efficiency of implementing the imported technology by WRF was low, in terms of both the delay of the technology transfer process and the overspending of investment funds. Because the management of the firm was poor and the utilization rate of imported equipment was low, the economic results of these two projects in terms of GVIO and profits were not satisfactory and the poor economic performance in turn led to a deterioration in the technological capability of the firm.

There are four reasons for the difference in the performance of the four firms in utilizing imported technology before the late 1980s. First of all, the overall firm management in terms of labour discipline, production planning and product quality control at CRF and WWF was better than at Westlake and WRF. Therefore, the first two grew relatively faster than the others. Second, the technological capability of CRF and WWF was relatively higher and these two entered their respective sector earlier than other firms. Consequently, they had the ability to put the imported equipment into production with less difficulty than the other firms. Third, the technology and equipment imported by CRF and WWF were more appropriate than those of the other two firms. Finally, because CRF and WWF performed better than other firms, they obtained more financial support from the government. Consequently, they established a comprehensive production capacity, which allowed them to have a technological edge over other state firms.

Declining economic performance and the deterioration of technological capability from the late 1980s

Although these four firms performed differently during the mid-1980s, their technological behaviour was similar. None of the four firms invested much in assimilating the imported technology. They mainly focused on the expansion of production capacity and did not try hard to reduce their production costs or to develop their new products to suit changes in market demand. Consequently, when the market became a buyer's one after the late 1980s, no matter how well they performed before, they all could not compete with other domestic firms at a lower price. Moreover, because they were basically passive users of imported technology and did not differentiate their products, though their facility was advanced, they could not compete with foreign firms in the high grade segment of the market.

The failure of these four firms to improve their technological capabilities during the late 1980s was mainly reflected in their inability to compete with non-state firms in the market. The achievement of CRF and WWF in establishment of a comprehensive production capacity during the mid-1980s was mainly due to government financial support and their privileged position as the first entrants in the protected tape recorder market. Because they did not make efforts to learn how to use the imported technology efficiently during the utilization period, their production costs did not fall as their cumulative output increased. Consequently, when the price of tape recorder products declined, when the market became highly competitive in the late 1980s, their high production costs prevented them from selling their products and led to a sharp decline in profits. In this respect, WWF is a good example of how state firms failed to take advantage of their advanced production facilities. Table 7.1 shows that WWF did not reduce its unit product costs quickly enough as its output increased from 2.8 million sets in 1987 to 4.2 million sets in 1989. Nevertheless, because the actual price of tape decks before 1989 was even higher than forecast, the firm still enjoyed reasonable unit profits. When the market price of tape decks declined quickly after 1989, the high unit costs prevented WWF from increasing its sales. As a result, it did not enjoy the economies of scale which it should have enjoyed if it had fully utilized its modern production capacity.

Since the late 1980s, the profitability of the four sample firms has declined

Table 7.1 Comparison between planned and actual unit costs of tape decks in WWF

	Unit prices (Yuan/unit)		*Unit costs (Yuan/unit)*		*Sale taxes (Yuan/unit)*		*Unit profits (Yuan/unit)*	
	Predicted	*Actual*	*Planned*	*Actual*	*Predicted*	*Actual*	*Predicted*	*Actual*
1984	42.00	41.46	35	35.68	2.1	2.26	4.9	3.52
1985	40.75	42.22	32.7	35.09	2.05	2.26	6	4.87
1986	40.50	43.33	32.6	34.55	2	2.03	5.9	6.75
1987	40.00	42.13	32.1	34.24	2	2.01	5.9	5.88
1988	39.00	40.8	31.25	33.63	1.95	1.96	5.8	5.21
1989	38.00	38.59	30.5	33.21	1.9	1.91	5.6	3.47
1990	37.50	36.45	30	33.15	1.9	1.88	5.6	1.42
1991	—	35.41	—	34.11	—	1.85	—	–0.55
1992	—	35.33	—	33.25	—	1.84	—	0.24
1993	—	35.05	—	32.57	—	1.92	—	0.56

Source: Own interviews.

dramatically, no matter how well they performed before. In 1991, three of these four firms suffered losses. Although CRF managed to remain profitable, the profits it made in 1991 were only 18.7 per cent of those in 1988. Faced with this difficult situation, the firms tried various measures to improve their performance. However, the performance of CRF and WWF recovered only slightly. The losses of the other two (Westlake and WRF) continued to increase. In 1993, their total debts exceeded their total assets. It seemed that there was no way to rescue them.

TWO COMPARATIVE CASES OF NON-STATE FIRMS

The declining profits of the four sample firms in the late 1980s was mainly due to their failure to reduce production costs to keep pace with the price declines when the market for tape recorder products became competitive. By contrast, FDI firms and TVEs began to enter the tape recorder industry in the late 1980s and developed quickly (see Chapter 4). The fact that state firms lost market share to these non-state firms reflected that the latter were more efficient in production and more technically adept. The following two comparative case studies show that firms with different institutional arrangements behaved differently from state enterprises, thus illustrating that the institutional arrangements governing the operation of firms played a major role in defining the technological behaviour and hence the performance of firms.

A comparison between CRF and WEA

Wujin Electronic Audio (WEA) is a township enterprise located in Hengtang Township of Wujin County which is under the jurisdiction of Canzhou

Municipality. This enterprise was originally set up in the early 1980s to do subcontracting work for state enterprises in the electronics industry. In 1986, WEA became a member firm of the Canzhou Electronics Industry Corporation (CEIC) which was actually an administrative body rather than a commercial company. The mission of CEIC was to control and promote the production of electronic products in the Canzhou area. Because WEA had long been a local supplier of CRF, CEIC asked CRF to diffuse its technology to WEA. The technical assistance provided by CRF included technological services and consultation in process engineering and industrial engineering, the training of technical personnel and workers, production planning and quality control. In addition, CRF relocated the production of several low grade tape recorders to WEA and provided all technical documents, moulds, non-standard tools, and key parts. It also allowed WEA to use the brand name of GLOBE to sell its products. All this was done under instructions from CEIC which, acting as a government agency, wished to enlarge the local revenue base by expanding the local output of tape recorders, which (as we have seen) had a relatively high profit margin during the mid-1980s. As it turned out, WEA soon became a large tape recorder producer and outperformed CRF in the early 1990s (Table 7.2).

Canzhou Electronics Industry Corporation was dismantled in 1989 because this type of administrative corporation was regarded as an obstacle to the autonomy of state enterprises. As the competition in the market for tape recorders increased, a dispute arose between these two tape recorder producers. The direct cause was WEA's use of the brand name of GLOBE, which was the registered trademark of CRF's products. Under the control of CEIC, WEA took advantage of this famous brand name for its products free of charge for more than five years until 1991, when CRF asked WEA to pay a royalty for its use of the brand name. In response, WEA stated in a national newspaper that from now on the brand name of GLOBE for WEA's tape recorders would be changed to NEWTECH. CRF thought that this state-

Table 7.2 Comparison between Canzhou Recorder Factory and Wujin Electronic Audio (1993)

	Canzhou Recorder Factory		*Wujin Electronic Audio*	
	Total	*Per employee*	*Total*	*Per employee*
Turnover (RMB ¥10,000)	22,833.2	7.64	14,676.5	8.08
Profits (RMB ¥10,000)	552.3	0.18	959.0	0.53
Fixed Assets (RMB ¥10,000)	6,125.3	2.05	2,136.0	1.18
No. of employees	2,987	—	1,816	—
Ratio of profits to net fixed assets (%)	9.02%	—	44.90%	—

Source: Own interviews with managers of CRF and WEA.

ment would create a false impression to the public and lead to a decline in the sales of its own products, which were still under the brand name of GLOBE. CRF tried to bring a lawsuit against WEA, but failed, because of the intervention of the local government which thought this dispute would damage the local tape recorder industry.

The dispute between CRF and WEA was seemingly over the use of the brand name. However, it revealed the different patterns of technological behaviour of firms under different institutional arrangements and the implications of different technological behaviour to enterprise performance as market conditions change. As was mentioned before, the sample state firms tended to use capital intensive technology. They were also restricted by the rigid labour system and other government regulations. Therefore, it was difficult for them to reduce production costs. By contrast, WEA, as an enterprise owned by the community of Hengtang township, was operating in an institutional environment different from that of state enterprises. As a TVE firm, WEA could only rely on its own resources. It could obtain bank loans from the Bank of Agriculture but at a higher interest rate than that available to CRF. Therefore, it tended to use labour intensive technology to cut the costs of fixed asset investment, because the cost of capital was relatively high for township enterprises. Moreover, as a township enterprise, WEA did not enjoy the monopolistic privileges and government protection that most state enterprises had. Therefore, it was more market oriented and intent on cost minimization. In order to cut costs, WEA used cheaper electronic components, simplified the design of the electric circuit of its tape recorders, paid less attention to how to meet the state standard of acoustic fidelity, and did not use expensive measuring and testing instruments. However, it tried to improve the appearance of its tape recorders, because its target market was the rural areas where the demand for tape recorders was increasing and the customers preferred good appearance to good tone quality. Therefore, these products could still be sold at a relatively high price, although the internal quality and the technological level of WEA's tape recorders were lower than those of CRF. This was one of the reasons for the dispute between these two firms, because CRF was proud of the quality of its own products and thought that the practices of WEA would ruin the goodwill of the famous GLOBE brand.

In addition to the fact that the capital goods used by WEA were relatively cheaper, labour costs were also much lower for WEA than for CRF. As a township enterprise, WEA's workers were farmers from rural areas.[1] Although the firm was nominally owned by the whole community, employment was based on a contract and not for the lifetime of the workers, as in state enterprises. Therefore, it was relatively easy for WEA to adjust its labour force according to its output, because the assembly of tape recorders was a relatively simple and routine job. As a result, the total labour productivity of WEA was higher, although it was less capital intensive than CRF (Table 7.2). Moreover, as a township enterprise, WEA did not need to meet such welfare expenses as housing and fringe benefits, and its wage level was relatively lower. This further contributed to its low production costs as a major competitive advantage.

Another important reason for the better performance of WEA is that WEA fully exploited the experience of CRF in tape recorder production. As a late user of the technology of tape recorder production, WEA learned from the successes and failures of CRF in adopting the imported technology. In addition, due to government intervention, WEA enjoyed extensive technological assistance from CRF, that would have been quite expensive to obtain if CRF had been a private enterprise. Consequently, WEA had taken advantage of a free ride of technology diffusion from state enterprises and adapted the learned technology to the institutional setting specific to TVEs, under which the labour costs to TVEs were much lower than those to state enterprises (see Jefferson and Rawski 1994: 61–2). Therefore, WEA, as a TVE, was able to compete with state enterprises and erode the high profits previously enjoyed by state enterprises, although the latter were technologically more advanced.

A comparison between WWF and Jianghai Electronics

Jianghai Electronics Co., a FDI firm located in Guangdong Province, is a joint venture between a Chinese foreign trading company (FTC), a local township enterprise in Zhuhai, Guangdong and a Hong Kong company. The major business of Jianghai is to assemble CKD kits of tape decks imported from Hong Kong and to re-export the final products to the world market through the distribution channels controlled by the Hong Kong company. Because the FTC was the export agent of WWF, when Jianghai was established in 1988, a vice director of WWF was recommended by the FTC as the general manager of the joint venture. Moreover, except the marketing department, a large part of the middle level technical and production management positions in Jianghai were staffed by the former employees of WWF, who were attracted by the relatively high wages and other benefits offered by the FDI firm. In this sense, therefore, the FDI firm transferred little technology to the host country except for the knowledge of how to organize production on a large scale. In 1993, with a total of 269 employees, Jianghai assembled 12.95 million sets of tape decks. Total sales were US $29.68 million, equivalent to RMB ¥264 million, and the total profits were RMB ¥5.1 million.[2] According to CAIA (1994), there were three tape deck assemblers in Guangdong province (including Jianghai) whose output each exceeded 10 million sets of tape decks.

Compared with Jianghai, there were several reasons for the persistently high labour costs and overhead costs of unit product in WWF.[3] First of all, the labour productivity of WWF was low (Table 7.3). In 1993, with 1,119 people, the tape deck division of WWF produced 3.75 million sets of tape decks, on which the turnover was RMB ¥131.44 million. By contrast, Jiang-

Table 7.3 Comparison between WWF and Jianghai (1993)

	WWF Tape Deck Division	*Jianghai Electronics*
Turnover per employee (RMB ¥1,000)	117.5	983.0
Turnover per assembly worker (RMB ¥1,000)	252.8	1,039.4
Output[a] per employee (1,000 sets)	3.4	48.1
Output per assembly worker (1,000 sets)	7.2	51.0
Profits per employee (RMB ¥1,000)	1.9	19.0

Source: Own interviews with managers of WWF and senior staff of CAIA; and CAIA (1994).
[a] Tape decks assembled by Jianghai were simpler than WWF's products. If the selling price is used as a measure, one WWF tape deck was equivalent to two Jianghai tape decks.

hai assembled 12.95 million sets of tape decks and the total turnover was RMB ¥264.3 million.[4] Therefore, the average turnover per employee of Jianghai was eight times that of WWF. If we take only the assembly activity into consideration, the average turnover per assembling worker of Jianghai was still four times as high as that of WWF.

Second, WWF was an 'all-around' firm. It possessed a comprehensive production capacity and produced most of the metal and plastic parts of tape decks by itself. Consequently, it did not enjoy the economies of scale. Unlike WWF, Jianghai specialized in assembling imported CKD kits for export. Moreover, the major capital goods in Jianghai were assembly lines which were relatively cheap. As a result, the unit capital costs of tape decks in Jianghai were much lower than those in WWF.

Third, as in all state owned enterprises, the workers of WWF had lifetime jobs. Their wage level in recent years had increased quickly. Moreover, they had many fringe benefits. Due to the low labour productivity of WWF, the proportion of unit labour costs in the unit value added of WWF was quite high. By contrast, Jianghai as a FDI firm located in the special economic zone was allowed to recruit its workers from rural areas on a contract basis (see Chapter 3). Therefore, although the nominal wage level was slightly higher in Jianghai, the labour costs per employee were similar for these two firms, for Jianghai did not provide fringe benefits, such as housing, for its employees. Because Jianghai could adjust its labour force according to the market demand and because contracted workers from rural areas worked harder than lifetime employees in urban areas, the labour productivity in Jianghai was higher. As a result, unit labour costs in Jianghai were much lower than those in WWF.

Fourth, WWF's unit overhead costs were much higher as well. This was due to the high indirect labour costs, the burden of retired workers and the high working capital costs in WWF. As a state enterprise, WWF had a large number of employees, who accounted for 10.7 per cent of total employees, working in the unproductive departments such as the party committee, propaganda department, canteen, kindergarten, trade union, clinics, etc.

Moreover, WWF had the responsibility to provide pensions for its retired workers whose number in 1993 was more than 700. Furthermore, because of the piling up of unmarketable products, the interests on bank loans as working capital generated high capital costs for WWF. By contrast, in Jianghai, there were only fifteen management personnel and the assembly process was labour intensive. Moreover, it did not have the problem of providing incomes for retired workers, because it was a new firm and its workers were quite young.

THE TECHNOLOGY RENOVATION PROGRAMME AND THE PROPENSITY TO USE CAPITAL GOODS IN THE SAMPLE FIRMS

The case studies show that during the mid-1980s, when they conducted technology imports projects, the four sample firms had the propensity to import more capital goods and did not emphasize the assimilation of imported technology. This is because they operated in a protected market and their financial resources to import foreign technology came mainly from the government. The case studies also show that their investment projects were subject to government intervention and control. Therefore, their technological behaviour was like a mixture of the 'engineering man' and 'bureaucratic man' approaches (see James 1989: Chapter 3).

All of these four firms conducted their technology import projects under the programme of renovating existing enterprises. As discussed in Chapter 3, because of economic reforms and the large number of projects in this programme, the decision making autonomy of local governments and enterprises in proposing and conducting technology imports were greatly enlarged. Because financial resources were highly subsidized and administratively allocated, state enterprises were motivated to put their projects into the programme by making their proposals consistent with the government policy and regulations. Because at that time the domestic market was highly protected, enterprises which produced goods to substitute for imports could easily earn high profits. Consequently, state enterprises preferred to use imported equipment to produce new products which other domestic firms could not produce without foreign technology and imported equipment. The cases of CRF and WWF show how state enterprises with privileges established their comprehensive production capacity for tape recorder products during the mid-1980s under the technology renovation programme.

The propensity to use capital intensive technology

Given the high profitability of tape recorder products during the mid-1980s, the easiest way for CRF and WWF to increase their own interests in terms of retained profits was to expand production capacity. Moreover, if the production expansion projects were included in the technology renovation programme within the plan of the government, financial resources would be cheaper to the firm.[5] In order to obtain subsidized investment funds, CRF and WWF turned to the establishment of a comprehensive technological capacity which appeared to meet the policy goals of importing technology to renovate existing enterprises and of localizing imported products.

For example, in order to localize its production of plastic parts, CRF's attempt to establish a mould making facility cost 65.5 per cent of the total investment funds of its second technology import project (Table 7.3). In late 1986, CRF signed a contract with IKO, a West German company that provided the technological advice for the manufacturing of plastic moulds. With the help of IKO, CRF imported and installed a number of mould processing machines and testing instruments in the first half of 1987. The machinery and instruments included milling machines, grinding machines, erosion machines, a computer aided manufacturing (CAM) centre, a horizontal production centre, a set of heat treatment facilities and a number of metallurgical material testing instruments. In addition, the firm imported a number of plastic parts decoration machines, such as facilities for vacuum coating, spray coating, surface treatment and gilding. It was planned by CRF that after the completion of the second project, it would be able to design and manufacture large and sophisticated plastic moulds with high precision. The planned mould manufacturing capacity and the planned output of plastic parts were 720 sets and 6.42 million pieces per annum, respectively.

Similarly, the focus of WWF in importing foreign technology was also on establishment of a modern production facility. In its feasibility study for the first technology project, WWF identified five major technical weaknesses which had to be overcome if it wanted to fulfil the target of producing one million sets of tape decks with homemade parts. These weaknesses were:

- The quality of homemade dies was poor, in terms of short service life and low precision of dies caused by the lack of skilled workers, high precision machine tools, measuring instruments, and design capability;
- The factory had no production capacity and experience in metal parts production on a large scale. In particular, it had no experience in production management and process engineering for production on a large scale;
- Design capability was poor, especially in the design of tape deck products, punching dies and non-standard tools;
- There was a lack of instruments to conduct the physical and chemical analysis of materials, processed parts and finished products to ensure product quality; and
- There was a lack of specialized machine tools and equipment, which prevented output from being increased and the quality from being guaranteed.

In order to remove these bottlenecks, in its three technology import projects WWF imported a large amount of capital goods. For instance, the specific target of the first project was to establish a basic parts production capacity of one million tape decks by importing a number of key pieces of equipment and machine tools for punched metal parts, axles, and mould-extruded plastic parts. These machines cost 93.2 per cent of the total investment of RMB ¥2.53 million during the first project.

WWF tried to establish a die/mould making facility in its second technology import project, because it could not have its high precision dies and moulds made by other domestic factories in this region. As a result, a large part of the total investment, especially foreign exchange, was spent on the purchase of instruments and machine tools, which accounted for 72.1 per cent of total investment and 96.4 per cent of total foreign exchange spending respectively.

Furthermore, CRF and WWF tried to meet all the technological specification requirements formulated by the governmental quality control departments. This is because, as state enterprises, they were subject to various government regulations. In addition, high quality standards and the technical advancedness of their products helped them to obtain more support from government agencies, especially from central industrial ministries. For example, both firms spent considerable amounts on the purchase of physical and chemical inspection instruments. This increased the production costs of these two firms, although their product quality and technological level were high as a result.

For example, in its second import project, CRF tried to modernize its product designing and quality control facilities. The steps taken were: (1) the importation and installation of a set of computer aided design (CAD) facilities and of a number of inspection and testing instruments for the quality control of electronic components and final products, and (2) the establishment of an acoustic fidelity laboratory. It was hoped that with the CAD facility, the process of developing a new product would be speeded up from nine months to three-and-half months and that with the advanced testing instruments the firm would be able to design high grade products. The specific goal of the firm was to establish a design capacity which could develop and launch seven to ten types of new products annually, as compared with only two to three new products during 1984/85. In its technology import projects, WWF also purchased many high precision and automatic machine tools which were seen by its managers as the key to the establishment of a comprehensive production capacity in a short time.

Government intervention

Both WRF and Westlake were subject to the control of government agencies in conducting their technology import projects, which, combined with poor management, led to the unsatisfactory performance in technology imports by these two firms. In the case of WRF, the specific type of technology of its first technology import project was chosen by GBCBI which wanted to establish a production capacity for plastic/metal tape decks. With the support of the central government agency, WRF's project was brought into line with the national technological renovation programme of machine building and electronics industries which consisted of 550 major projects in the National Sixth Five-Year Plan (1981–5). WRF's project was given a high priority in terms of financial resources, foreign exchange and material supply by both the central and provincial agencies. However, in spite of the priority, the whole process of approving, planning and implementing the project proposal was lengthy and troublesome. To get approvals and funds, WRF had to submit various documents to various departments at both the provincial and central levels,[6] which took nearly a whole year to complete. In addition, WRF had to obtain approval from various governmental organs. Moreover, different departments gave different instructions that sometimes conflicted with each other. For example, GBCBI stressed the technological attributes and the Economic Commission wanted to control the costs. WRF had to find a compromise between them. Consequently, the NTP tape deck was chosen and the original design was changed to keep the costs within the allocated budget.

In fact, WRF did not have the absorptive capacity to master the technology. The process technology of NTP parts, especially the plastic/metal combined parts, was too sophisticated for WRF to fully master. However, in order to have its technology import project brought into the state plan, WRF imported the technology selected by GBCBI. When its project was approved and allocated funds, WRF imported fifty-six items of equipment and instruments, which accounted for 71 per cent of the total foreign exchange spending.

In 1986, when WRF was unable to manufacture this type of tape deck, the government compensated it by providing a bank loan to allow it to conduct another technology import project. This time the firm was even more biased towards imports of capital goods. It imported many pieces of equipment, including high speed auto punches, a hot-cabin casting press, and die-making facilities. The costs of equipment accounted for 93.5 per cent of the total foreign exchange spending (Table 6.8).

In the case of Westlake, the government agencies did not give priority to the tape recorder sector at first. Later on, when tape recorder production was separated from television production, the government allocated the funds evenly between the two firms. Each of them was limited to final product assembling. Consequently, when these two firms merged into one firm, its part production capability was quite poor.

THE CMRS SYSTEM AND THE LACK OF ASSIMILATION ACTIVITIES

As state enterprises, these four sample firms implemented all of the major reform measures, including the loans-for-grants in investment financing and the tax-for-profits in profit sharing with the government. In 1987/1988, these four firms began to adopt the Contract Management Responsibility System (CMRS). The practice of CMRS in these four sample firms shows that the contract system introduced as a reform measure did not encourage firms to invest in technological activities because it did not solve the problems caused by ambiguous property rights to the assets of state firms. The purpose of CMRS was to encourage greater efficiency in state firms by allowing them to keep more when they generated higher profits and to guarantee government taxation revenue (see Chapter 3). It was hoped by the authorities that the behaviour of state enterprises could be regulated through the terms of the contract. This system actually institutionalized the property rights that the government, the firms and the employees of state firms each had to the industrial assets. However, the property rights of each party were neither clearly delineated nor enforceable in terms of the exclusivity and transferability of these rights. Consequently, the contract between the government and state firms could not solve the problem of common property, and the contract could not be effectively enforced. As a result, each party would be expected to try to invest less in, and capture more from, the resources (Cheung 1986; Wu Jinglian 1993b; Lee 1991, and Jensen and Meckling 1979).

The short term behaviour under CMRS

In particular, the CMRS was not effective in regulating the technological behaviour of firms, because of the nature of technology as an intangible industrial asset. It was very difficult to stipulate in and to enforce by the form of contract the improvement of technological capabilities. Therefore, in the CMRS contracts, the obligation of the four firms in enhancing their technological capability was only stipulated in terms of the technological investment that they had to make with their retained profits. For instance, in the period of 1987 to 1988, when CRF was carrying out the second technology import project, this project automatically became the technological part of the CMRS contract. However, for the following two years of 1989 and 1990, there was no specific requirement regarding technological advancement in the contract. Instead, the contract simply stipulated that the net fixed assets of CRF should increase by 5 per cent annually during the contract period and that CRF should invest no less than 37.5 per cent of its retained profits in technological projects.

The key factor in the CMRS was the distribution of income between the government as the nominal owner of state firms and the firm as a collection of its employees. When drawing up the contract in the first run of the CMRS, the contract terms regarding the total realized profits, the sharing of profits between the government and the firms, the income of workers, and the obligation of the firms to invest in technological activities were easily worked out based on previous performance, since both parties were optimistic about the market for tape recorders and the government allowed the firms to keep a large part of the profits above the contracted target. For example, in its contract, CRF was allowed to retain two-thirds of the increment, if it could over-fulfil the target of total realized profits, provided it used the incremental funds for technological renovation projects. If CRF could not meet the profits target, it would have to use its own funds to cover the shortfall. In the contracts of the other three firms, there were similar clauses.

Encouraged by the prospect of retaining more profits, the firms tried to expand their output. For example, WWF invested its retained profits during the period 1988–1989 to remove the bottlenecks which emerged in achieving the output target of the third technology import project, by adding some pieces of equipment and instruments. In other words, these technological activities, instead of being aimed at increasing the efficiency in using the imported technology and/or improving its technological capability to respond to changes in market conditions, were aimed at increasing the output from 2.8 million sets in 1987 to 5.2 million sets in 1990.

Similarly, Westlake also tried to maximize its short term cash flow through expansion of its existing products. During the period 1988–1989, the firm purchased a number of plastic extruders to replace its original outdated machines. Moreover, it greatly increased the output of its existing products by stretching its production capacity, despite insufficient orders from its customers, because the firm's executive managers thought that they could sell those tape recorders sooner or later. But, because the major products of Westlake were still dual deck AM/FM stereo portable tape recorders, which became outdated in the late 1980s, the firm could not sell its tape recorders in the market. In 1990, Westlake incurred losses. Its superior, the Bureau of Instrument-making Industry, tried to help it. It was stipulated in the contract of the second run CMRS, starting from 1991, that the reduction of loss of the firm as compared with the performance in the previous year could be regarded as profits. In other words, Westlake could keep a certain percentage of any reduction in losses as a subsidy. Nevertheless, because Westlake did not establish competent technological capabilities, its performance failed to improve.

The tendency to maximize employee income and the lack of technological initiative

In the CMRS contracts, the wage, bonus and employee welfare benefits were closely linked to the total profits of enterprises. The government hoped to regulate the behaviour of firms by stipulating the proportions of bonus, welfare payments and technological investment in the firms' retained profits. However, due to the ambiguity of property rights of the government, the firm and workers to the industrial assets, the wage and bonus schemes in the CMRS system was difficult to enforce. In reality, the wages and bonuses of employees of the sample firms tended to grow faster than the profits. Moreover, when the firms performed poorly, they tended to bargain with the government for reductions in profit remittance and in investments in technology. The high inflation rate from the late 1980s certainly contributed to the fast growth of employees' incomes in the firms.[7] However, as illustrated by the example of CRF, as long as they were still profitable, the income of SOEs' employees tended to increase at a rate higher than the inflation rate. The income increase was sustained by sacrificing investment in technology.[8]

In its CMRS contract, CRF adopted the widely used wage and bonus scheme: the increase in its total wage bill was linked to the growth of its total realized profits. The proportional link of the two rates of growth was 1 to 0.7, that is to say, with a 1 per cent increase in the realized profits, the total wage bill of CRF was allowed to increase by 0.7 per cent. When the scheme was implemented, a total wage bill, after complicated calculations and negotiations, was worked out as the base on which the total wage bill for each year during the contract period would be determined according to the proportional link. Under the terms of the contract, the total bonus of employees was stipulated to be less than 32.5 per cent of its retained profits.

As the market became more competitive in the late 1980s, the profits of CRF decreased dramatically from RMB ¥17.3 million in 1987 to between RMB ¥3 to 5 million in the early 1990s. In spite of the deterioration in profits, the wage level of employees increased faster than stipulated in the contract. In reality, the annual growth rates of average monetary incomes of CRF's employees were 17.3 per cent and 11.9 per cent during the periods of 1987–90 and 1990–3 respectively (Table 7.4).

When the business performance of CRF deteriorated, instead of economizing in its operations, CRF began to bargain successfully with the government for the reduction of remitted profits. The government agreed to reduce the profit remittance targets of CRF in the second run CMRS. This is because it was difficult for the government to find out whether bad management of CRF or changing market conditions was the real reason for the poor

Table 7.4 Average annual wage per worker in CRF

	RMB ¥/per capita[a]
1987	1,873
1990	3,021
1993	4,234
Annual growth rate	%
1987–90	17.3
1990–3	11.9
1987–93	14.5

Source: Own interviews.
[a] The income figures are at current prices. The average annual inflation rates were 12.4%, 9.8% and 11.1% during the periods of 1987–90, 1990–3 and 1987–93 respectively (SSB 1994: 213).
Note: The average income of worker is the total wage bill divided by the number of workers. It does not include fringe benefits in kind.

performance of the firm, and because the government did not have the option to contract with others (Wu Jinglian 1992: 165–72). Although in the contract there was a clause stipulating that, if CRF could not fulfil the target of total profits, the gap should be met with the firm's own funds, this was never seriously enforced. This is because the firm under the current institutional arrangement was not an independent entity possessing its own property which could be used to compensate the government for its losses.

In the second run of the CMRS, CRF proposed to carry out an additional project to develop hi-fi stereo systems, because through the production of high grade audio products it could exploit the existing advanced technological facilities which were its comparative advantage in competition with township enterprises. However, it failed to obtain the necessary bank loans. The government lacked confidence in the ability of CRF to compete with foreign products which dominated the domestic market for high grade audio products and which were much more advanced than those that CRF could make. Furthermore, due to the short term behaviour problem and common property problem (see the case of CRF and Westlake under the CMRS discussed in this section), it was unlikely for the firm as a collection of employees to invest by sacrificing current income in long term projects to which the property rights were ambiguous. As a result, CRF did not make sufficient investment to up-date its technology and its net fixed assets decreased in 1992 and 1993. Because it had not improved its technological capability and was still stuck in the production of its old products, on which the profit margin was shrinking quickly, its performance in terms of total realized profits failed to recover.

CONCLUDING REMARKS: THE INSTITUTIONAL ENVIRONMENT AND THE ROLE OF MARKET COMPETITION

The purpose of the case studies is not to demonstrate rigorously the relationship between institutional arrangements and firms' technological behaviour and performance. Rather, the case studies illustrate the impact of institutional arrangements on the process of technology transfer in state firms by showing how state firms behaved in a given context. They illustrate the circumstances in which state firms operated, the activities they undertook during different phases of economic reforms and the outcomes of their activities. These case studies supplement the general survey with vivid details to help readers to understand the whole process through which institutional changes affected the technological behaviour of these four firms.

The case studies show that the performance of technology imports in the four sample state enterprises was unsatisfactory, in terms of both technological capability enhancement and the efficiency in using imported technology and equipment. If assessed against the analytical framework, CRF and WWF once reached the 'higher level' of assimilation. However, this was achieved with the funds subsidized by the government and in a protected market. The other two firms, Westlake and WRF, remained at the 'lower level' of assimilation. That is, their production activities were basically limited to assembling imported CKD kits, operating imported machinery and imitating the imported products.

Although these four firms performed differently during the mid-1980s, they all did not invest much in assimilating the imported technology and their technological capabilities in using imported equipment efficiently and in developing new products quickly were weak. Furthermore, because they did not make efforts to modify both the imported process technology and product technology to adapt them to local conditions, they failed to establish the technological capability to improve the imported technology and innovate based on the assimilated foreign technology.

The quick maturity of the tape recorder market caused by the duplicated imports of tape recorder production facilities and the entry of more non-state firms pushed down the price of tape recorders (see Chapter 4). Faced with the competitive market, the economic performance of the sample firms declined dramatically. At the very least, this is consistent with the proposition that the technological capability of the sample firms based on the passive use of capital goods was inappropriate in a competitive environment. Consequently, the poor performance of the sample firms in market competition led to a deterioration in the technological capabilities of these firms.

The experience of these firms shows that the impact of institutional arrangements on the process of technology transfer is of primary importance. This is because how firms respond to the microeconomic and microtechnical factors is shaped and constrained by the institutional arrangements governing their operation. Therefore, when the institutional arrangement remains the same, micro factors, such as the type of technology chosen, the original technological capability of the recipient firms and the level of enterprise management, may result in different levels of performance, as shown by the different performance of CRF and WWF as compared to that of Westlake and WRF. When the institutional arrangement is different, however, its impact on performance becomes significant. This is illustrated by the declining performance of these sample state firms when they were challenged by non-state firms in the market after the late 1980s.

The bias of these firms towards capital goods in importing technology in the mid-1980s can be attributed to the macroeconomic factors of the privileged position of state firms and the protected market. Because of the subsidized funds under the programme of renovating existing enterprises and the lack of market pressure, they tended to use capital intensive technology and to neglect the assimilation of imported technology to use the imported equipment more efficiently and to update the imported products more quickly.

The CMRS system further blurred the property rights structure in state firms. Under this arrangement they tended to maximize the short term income of their employees by sacrificing the long term development of their technological capability. The case studies demonstrate how managers and workers in state enterprises under the institutional arrangement of CMRS tried to squeeze their near term benefits from the industrial assets by reducing investment in technology because the industrial assets of state enterprises were becoming a sort of common property. In particular, technology as an intangible industrial asset was actually put in the public domain, because it is difficult to use the contract management responsibility system to stipulate the attributes of technology and the responsibility of a firm as a collection of its total employees for the development of technology. Consequently, the rent from the resources of the industrial assets in state firms under the CMRS was often dissipated (Cheung 1974, 1986; Jensen and Meckling 1979; Wu Jinglian 1993a).

The poor performance of the four firms and the resultant deterioration of their technological capability in the early 1990s suggests that market competition alone could not automatically regulate the behaviour of state enterprises, if the property rights to the assets were not clearly and exclusively defined. It also illustrates that an inconsistency in the institutional framework and policies prevents firms from re-allocating their resources efficiently. On the one hand, the operational environment of state firms changed very much as a result of economic reforms. The product market became competitive and state firms were asked to be responsible for their own performance. On the other hand, a number of fundamental institutions remained basically the same as under central planning. State firms were still obliged to provide incomes for their lifetime employees. The ownership of state firms remained ambiguous and was still exercised by governmental agencies. Under this situation, state firms not only could not cut their production costs by dismissing superfluous workers, but also could not re-allocate their assets to other valuable uses when facing operational difficulties. Consequently, the technological capability of these firms could not be maintained, let alone enhanced, because of the nature of technology, in particular, indivisibility and depreciability (Nelson 1959).

8 Conclusion

This study has analysed the impact of institutional arrangements on the technological behaviour and the performance of Chinese state owned enterprises (SOEs) in transferring foreign technology during reforms. The empirical results of this study have shown that, although Chinese SOEs were granted enlarged decision making autonomy in their daily operation and the market became increasingly competitive as a result of reforms, they were still not motivated to learn more intensively about imported technology. Rather, due to the continued absence of well delineated property rights to industrial assets in which technology constitutes an important part, SOEs tended to maximize their short term 'rents' from their imported technology without investing resources to absorb and improve it. Consequently, the technological capabilities of the firms deteriorated quickly.

THE TECHNOLOGICAL BEHAVIOUR OF SOEs IN IMPORTING AND UTILIZING FOREIGN TECHNOLOGY

A general evaluation of technology transfer by the sample SOEs

As has been pointed out in Chapters 2 and 4, effective technology transfer at the firm level in this study is defined and evaluated along two main dimensions: the improvement in technological capabilities of the recipient firms and their overall change in economic performance as a consequence of importing and utilizing foreign technology.

Generally speaking, if we use the technological tasks the recipient firms undertook as a proxy to evaluate the effects of technology transfer (see Mytelka 1985: 81), the sample state firms gained basic technological capabilities in tape recorder production as the result of technology import projects (see Chapter 4). That is to say, most sample state firms had gained the ability to implement and operate the imported technology and production facilities, without changing them substantially. In order to use locally available materials, some of the sample firms also modified the imported technology to adapt it to local conditions. A few of them went as far as to modify the imported technology significantly and developed new products by themselves. Therefore, if we assess the technological capability improvement of the sample firms in the second half of the 1980s against the analytical framework developed in Chapter 2, a large proportion of the sample firms did not enhance their technological capability beyond the 'lower level' of technology assimilation (see Figure 2.1). They remained at the phase of operating the imported technology and equipment in its original form. Nevertheless, a few firms achieved the 'higher level' of technology assimilation, in the sense that as the result of technology imports they established a modern production capacity involving new product development and parts production.

The technological capabilities in tape recorder production of the sample firms, though uneven, were achieved with government financial support and in a seller's market during the mid-1980s (see Chapters 5 and 7). Because their import projects were in the government plan, the sample firms had the privilege to enter the tape recorder market much earlier than other firms. Consequently, although most of the sample firms performed poorly in terms of their efficiency in implementing and utilizing the imported technology and equipment, they were still highly profitable. Because of the lack of competition, they were able to continue to produce and develop. Under this situation, a few of the sample firms with better management and/or a higher level of initial technological capabilities performed better than others. Therefore, they were allocated more government-subsidized funds which allowed them to import more advanced technology and production facilities (see Chapter 6).

Nevertheless, although they performed differently, the technological behaviour of the sample firms was similar. In general, they concentrated on purchasing capital goods when they carried out their technology projects. They focused mainly on expansion of production capacity in the follow-up projects and did not try hard to reduce production costs or develop their new products to suit changes in market demand. Consequently, when the market for tape recorders became a buyer's one from the late 1980s, they could not compete with other more efficient domestic firms that charged lower prices. Because they were basically passive users of imported technology and copied the foreign products which were outdated, the SOEs were unable to establish their own niches in the market for differentiated or high-grade audio products.

As far as the overall economic outcome is concerned, the performance of the sample SOEs was not satisfactory (see Chapter 4). Before the late 1980s, the sample firms mostly did not meet the planned targets of their technology import projects, although most of them were still profitable. They also failed to improve their production efficiency as their cumulative output increased. As market competition became increasingly fierce after the 1980s, all of the sample firms began to suffer from declining performance, no matter how well they had performed before. Due to their poor performance in market competition, many sample firms reduced their output of tape recorder products. As a result, their technological capability began to deteriorate.

The changes in the performance of the sample firms in utilizing imported technology demonstrates that the impact of such macroeconomic factors as institutional arrangements, government policy and market competition on the technological behaviour and thus the performance of firms in transferring technology was of primary importance. This is because the response of firms to micro factors was shaped and constrained by macroeconomic factors. During the mid-1980s when the tape recorder market was dominated by state enterprises, they enjoyed a high profitability. At that time, their technological behaviour was similar: preferring further capital goods purchases and lacking investment in technology assimilation. There was variation in the performance among these sample firms, but it was mainly determined by micro factors. When the market for tape recorders became competitive, however, as the result of the entry of more non-state firms into this market after the late 1980s, the performance of the sample SOEs, including those which had established modern production facilities, declined substantially, showing that they as a group under the same institutional arrangements were unable to compete with other firms under different institutional arrangements.

The agency problem and the propensity of SOEs to import capital intensive technology

The empirical investigation found that when undertaking import projects, the sample enterprises preferred such capital goods as assembly lines and machinery to licensed technology and technical know-how. They also tended to use advanced technology. Because of the high production costs and the inflexibility, the production capacity of advanced equipment established by the sample firms turned out to be inappropriate, when the market for tape recorder products became competitive. This kind of technological behaviour had the main characteristics of 'an engineering man' and 'a bureaucratic man' as discussed by James (1989). This is because the governance of China's SOEs was still through a hierarchical bureaucratic system and market competition was still lacking.

Nevertheless, the technological behaviour of China's SOEs during the mid-1980s was different from that of the traditional state enterprises. In the case of the sample state tape recorder enterprises, the agency problem in importing technology at that time was serious mainly due to three macroeconomic factors. They were (1) the enterprise reform which granted more decision making autonomy to SOEs; (2) the protected market for tape recorder products; and (3) the government sponsored programme of renovating existing enterprises.

As the result of economic reform, SOEs were allowed to retain a certain proportion of the profits they generated. And, because of the separation of 'management' and 'ownership', the management of SOEs had more decision making autonomy in their daily operation. Consequently, the objectives of state enterprises changed from seeking an easy way in fulfilling the targets given by the plan to seeking their own benefits in terms of the incomes of their employees.

The market for tape recorder products in the mid-1980s was not competitive because the domestic market was protected by the government from foreign products and the demand was far larger than the supply. Consequently, the production of tape recorder products was very lucrative at that time.

Most of the technology import projects conducted by the sample SOEs were under the programme of renovating existing enterprises with imported technology. In order to encourage enterprises to carry out import projects and facilitate the process of substituting imported products with domestic products, enterprises were given many privileges once their technology projects were brought into line with the programme. Because of the huge number of projects under this programme and the enlargement of decision making autonomy in SOEs, enterprises were given more discretion in proposing import projects and selecting the types of technology to be imported. However, because the programme was planned and implemented through the hierarchical system for central planning, the sample SOEs were still subject to the control of governmental agencies.

Because of these factors, the incentive structure was distorted. The sample SOEs tended to use capital intensive technology, as long as their projects were in the plan. At first, they preferred to import assembly lines and CKD kits because these would generate quick returns. Later on, when the government restricted the imports of assembly lines, they turned to equipment for manufacturing parts to enjoy the preferential treatment given to the localization programme.

The empirical evidence also shows that in carrying out their technology import projects, the sample firms tended to build up an 'all-round' production capacity with advanced equipment and instruments. There were three reasons for this kind of technological behaviour. First, it reflected the attitude of SOE managers and government officials towards the concept of technology. In their thinking, technology was something embodied in advanced machinery. In the selection of specific types of equipment, the costs were normally subordinate to the consideration of advancedness of equipment (McGuinness, *et al.* 1991; Conroy 1992: 231). Second, China's SOEs tended to carry out as many production tasks as possible for themselves, because the market mechanism to coordinate the production of specialized firms did not exist and the planning system was very primitive in China (Byrd 1987: 220; World Bank 1990: 171–4). Advanced and comprehensive production facilities would allow them to avoid asking other firms for help and thus protect their positions in the planning system. Third, one of the important policy objectives for importing technology was to increase the localization rate in the production of inputs. Firms which carried out localization projects, that is establishment of parts production facilities in the case of tape recorder sector, were provided with government financial support.

Compared with non-state firms, the sample state enterprises placed more emphases on the technical standards required by the techno-bureaucrats in the ministry. They used high quality electrical components and advanced manufacturing equipment and inspection instruments to meet those standards. This is because, as state enterprises, they were more subject to the control of the government and to the availability of government financing which were at least partly related to whether they could meet the technical and quality standards required by the government agencies, such as the signal–noise ratio and acoustic fidelity. However, the higher internal quality of the products of SOEs did not have much value in market competition because the technical standards were determined by the government agency and were not appreciated by the consumers (Langlois and Robertson 1995: 87). Furthermore, although they paid attention to the standards required by the government, they normally ignored or did not try to develop their new products with the high quality needed to meet the changing demands of the consumers.

The propensity of sample firms to use capital intensive technology was not consistent with the government policy which aimed to improve the effectiveness of technology imports and to reduce the costs of technology imports (see Conroy 1992: Chapter 5). This deviation reflected the agency problem of state enterprises in undertaking technology imports. The agency problem was striking during the middle 1980s. This is because the governance of SOEs was still through the hierarchical system designed for central planning (Jefferson and Rawski, 1994); the separation of 'management' and 'ownership' provided unprecedented incentives and opportunities for SOE managers to pursue their own benefits; the financial resources of technology import projects were subsidized and administratively allocated; and the pressure from the market for cost minimization was weak.

Because of the subsidized financial resources and other privileges together with the seller's market, the sample SOEs tended to pay less attention to the capital costs and the feasibility of their projects. This led to the low efficiency of many sample firms in implementing the imported technology. Furthermore, their production costs were persistently high, because the low utilization rate of the imported equipment prevented them from enjoying the economies of scale which capital intensive technology normally should have (Nelson 1994: 26). All of these factors led to the difficulties of the sample state firms in assimilating imported technology and the subsequent poor performance of the sample state enterprises in market competition.

The common property problem and the lack of learning efforts in SOEs

The process of technology transfer in the sample state enterprises entered the utilization and assimilation phases in the late 1980s and afterwards. During this period, the technological behaviour of the sample SOEs can be characterized by the common property problem. When the production of tape recorders was still profitable, the sample SOEs tended to maximize their short term incomes without investing in long term technology development. However, when market competition became increasingly severe, the economic performance of the sample SOEs declined dramatically and their technological capabilities deteriorated quickly.

The CMRS system and the short term behaviour of SOEs

In 1987/1988, the sample SOEs began to implement the CMRS system. The aim of the CMRS system was to encourage greater efficiency in SOEs while regulating their behaviour through the form of their contracts. This system actually institutionalized that both the government and firms possessed property rights to the industrial assets of state owned enterprises. However, the property rights of each party were not well defined in the sense that these rights were neither exclusive nor transferable. Consequently, the contract between the government agencies and SOEs could not be effectively enforced. As a result, each party could be expected to try to invest less in and capture more benefits from the resources committed (Cheung 1986: 66–73; Wu Jinglian 1993a; Lee 1992; and Jensen and Meckling 1979).

Under the CMRS system, the incomes of the employees of SOEs were linked to the profits that their firms generated. Encouraged by the prospect of obtaining more profits if they could over-fulfil the profit targets, during the first two years of the first run of the CMRS (1987–90), when the profit margins on their tape recorder products were still high, the sample SOEs tried to expand their output by purchasing additional production equipment and stretching their existing resources. The long term development of their technological capability was ignored in the sense that their investment was not aimed at improving and updating their technological capabilities, because this would not increase their short term cash flow.

The quick maturation of the tape recorder market after the late 1980s, which was caused by the duplicated imports of production facilities and the entry of more non-state firms, pushed down the profit margin on tape recorder products. Due to their persistently high production costs, the profit performance of the sample state enterprises declined dramatically. Faced with their poor profit performance, the sample state enterprises began to bargain with their supervising government agencies for lower profit remittances as well as reducing their investment in technological activities. In the meantime, they tried to keep the incomes of their employees increasing. This is because, under the CMRS system, workers had a *de facto* property right to claim incomes from the industrial assets while managers of SOEs did not have sufficient rights and incentives to resist the pressures of workers for income increases (Walder 1989; Huang 1990). Because pressures for increases in employees' incomes were much greater than contractual commitments for technology investment, when profits declined, the latter was the first to be sacrificed.

The lack of efforts to improve imported technology from the side of SOEs was a manifestation of the common property problem under the CMRS system. The CMRS system aimed to reduce the residual loss (as defined by Jensen and Meckling 1976) by linking firms' interests (in terms of the incomes of their employees) to their profit performance. Nevertheless, because of the absence of well delineated property rights, the incentive was weak for SOEs to re-invest their retained profits to improve their technological capabilities where the property rights to the technology did not belong to them. The government tried to stipulate in the CMRS contract the obligation of firms to enhance their technological capabilities. Due, however, to the inherent difficulties in using the responsibility system to define property rights to industrial assets (Cheung 1986; Jensen and Meckling 1979), as shown in the case of the sample firms, this system did not motivate SOEs to make efforts at long term technological development.

Furthermore, the CMRS system was implemented in a hierarchical system of state ownership. The government remained the owner of SOEs. However, under this arrangement, SOEs as the collection of lifetime employees were actually given the rights to use the assets and to share the profits from the assets. Consequently, it was very difficult for the government as the principal to monitor the performance of firms in general, and to ensure that firms would make technological efforts in particular. First of all, using profits as the major performance indicator induced moral hazard in firms as the agent. Firms would try to maximize their short term cash flow through increasing current profits by ignoring other aspects of their assignment, such as long term development, which would be beneficial to the principal. Second, the CMRS contract was not voluntarily signed between firms and the government agencies, due to the 'cellular' system of government ownership. Therefore, this arrangement prevented competition, which is the major mechanism to reduce the agency costs caused by separation of management and ownership (Jensen and Meckling 1976; Eggertsson 1990). Third, under the CMRS system, the profit targets were negotiable and firms did not have their own resources to compensate the principal in case their actions harmed the interests of the principal. Therefore, the opportunistic behaviour of firms could not be effectively controlled.

Institutional constraints and the deterioration of technological capabilities of SOEs

The empirical evidence showed that the economic performance of the sample state enterprises declined dramatically after the late 1980s when the market for tape recorders became increasingly competitive due to the entry of more firms into the market. As a result, the financial resources available to invest in technological activities were further reduced.

The deterioration of the technological capabilities of the sample state firms was aggravated by the institutional constraints binding the behaviour of SOEs. First, due to the fact that SOEs were owned by government agencies at various levels, the industrial assets of SOEs were not transferable through the market. Consequently, re-allocation of production factors was very difficult and of low efficiency. In this case, the imported production facilities of the sample state enterprises were mostly allowed to lay idle and lost their value quickly. Second, as the result of personnel reform, employees of SOEs had much more freedom and greater chance to choose their job. However, they still had the right to claim incomes from state enterprises. Consequently, there was a reverse selection effect. Well educated and experienced employees tended to leave the poorly performing state enterprises to find jobs in TVEs and FDI firms which were willing to pay a higher wage, while aged and less capable workers remained at SOEs. Finally, state enterprises had an obligation to provide income and welfare for their lifetime employees (Harrold and Lall 1993: 46–8). Faced with the decreased market share for their tape recorder products, many sample firms closed a large part of their tape recorder production lines to reduce costs and carried out various sideline businesses, mainly in tertiary industry, to earn income for their employees. This meant that the technological capability embodied in the firms and in human capital established as a result of the technology import projects in the past was no longer utilized, and therefore was being lost.

The deterioration of technological capability in the sample state enterprises shows that the competition in the market for tape recorder products and the hardening of budget constraints were not sufficient to regulate the technological behaviour of state enterprises. Because of the continued absence of well defined property rights to the industrial assets, SOEs tended to 'eat up' the assets rather than invest their retained funds in technology development. Therefore, it can be concluded that the ambiguity in the property rights structure was the major factor responsible for the technological behaviour of the sample state enterprises and their performance in transferring foreign technology.

THEORETICAL IMPLICATIONS

Technological capability improvement and active learning efforts

Historical experience shows that technology transfer is very important in the catch-up process of developing countries and that the critical problem is getting imported technology to work effectively and efficiently (Rosenberg and Frischtak 1985). Many economists have also pointed out that deliberate and extensive local efforts in assimilating imported technology are essential to successful technology transfer, in terms of long term technological capability development and improvement of economic efficiency (Lall 1987; Nelson 1990a; Pavitt 1985; Stewart, 1990; and Dosi and Orsenigo 1994). This research, using empirical evidence, demonstrates the importance of active learning in the establishment and improvement of technological capabilities by the recipient firms and the role of market competition in directing, motivating and evaluating the learning efforts of the recipients.

The importance of learning and the role of market competition in guiding learning efforts are closely related to the nature of technology. In modern society, technology as human knowledge applied to production has two different aspects: codified and generic information, and uncodified and firm-specific knowledge (Nelson 1987: 76; 1990b: 197; Pavitt 1985: 5, Mowery and Rosenberg 1989: 142–3; and Dosi 1988: 224). Although the codified and generic information embodied in capital goods and technical documents can be transferred through various channels, the tacit and firm-specific knowledge can only be learned through practice.

Advanced technology in terms of new product designs and more efficient processing equipment will greatly increase the productivity of the recipient firms. Even passive use of the imported technology will allow the recipient firms to use fewer inputs to produce more output, as compared with their previous performance with outdated technology. As long as they have the necessary absorptive capacity, the recipient firms will be able to install and operate the imported production facilities and to copy the imported products (Cohen and Levinthal 1990). This is because technology has certain properties of public goods (Nelson 1987: 74; Arrow 1962; Gomulka 1990: 25). Firms can draw benefits from the newly acquired codified technology, if they can get access to it. This was true in the case of the sample state firms when they imported the analog sound recording cassette tape recorder production technology which was mature and largely codified but was new to China in the first half of the 1980s (Baert *et al.* ed. 1988: 4–7). As they already possessed relevant knowledge in radio production, they were able to implement the tape recorder technology and to gain benefits from it.

However, understanding the generic aspect of technology does not in itself enable the recipients to use the imported technology efficiently, in the sense that given the imported technology the recipient firms could achieve the physically possible output. This is because efficiency in using a technology depends on how the recipients adapt it to the specific conditions under which they operate. That is to say that efficiency in using the codified technology depends on the mastery by the recipient firms of the tacit aspect of technology and that the latter is accumulated through active learning (Nelson 1990b).

Furthermore, the ability of recipient firms to improve the imported technology also depends on learning and accumulating the firm-specific and experience-based aspect of technology. This is because the improvement of a technology means the evolution of a technological paradigm to suit the changes in the market environment, to which there is no given codified solution (Dosi 1982 and 1984). Therefore, accumulated firm-specific knowledge is often the basis for firms to innovate and to create new technology (Rosenberg 1976: 76). In this sense, using one type of imported codified technology passively, without making efforts to learn how to adapt it to the specific local conditions, will not allow the recipient firms to improve their acquired technology. When this type of technology is obsolete and replaced by a new generation of technology, the recipients have to import the new type of technology again and the resources spent on the outmoded technology will be wasted. But, if the sample firms had developed their technological capabilities into the 'higher level' of technology assimilation, as we defined in the analytical framework, they would have the ability to keep pace with the development of this type of technology.

Because technology is mainly differentiated, dispersed and ever-changing knowledge (Pavitt 1985), market competition is the most efficient mechanism to direct and evaluate the learning efforts of the recipient firms (Hayek 1978/1984; Nelson 1990b). As is shown by the empirical evidence of this study, lack of market competition discouraged the sample SOEs from choosing the more appropriate types of technology when they carried out their projects. Due to the absence of competition, they had high profits at first. This distorted signal led them to invest in output expansion and in meeting the technical standards set up by the central planners rather than in satisfying the customers. Therefore, the technological capability established by the sample firms mainly stressed the codified aspect of the imported technology. Consequently, efficiency in using and efforts to improve the imported technology were unlikely. Furthermore, the distorted signal triggered the duplicated imports and the over-supply. Some people may argue that over-supply could increase competitiveness (Stoneman and Diederen 1994). However, duplicated imports caused a large amount of resources to be wasted. Because the imported facilities were similar and many firms did not make active learning efforts, this waste did not result in the emergence of a more advanced technology as the rivalry in a market economy normally does (Nelson 1990b).

Technological capability is ultimately evaluated and selected by market competition. In the case of China's tape recorder industry, many state enterprises with more advanced machinery and more engineers lost out in their competition with TVEs and FDI firms. This is because their technological capability was established in the absence of market competition, and, hence, ignored the accumulation of the firm-specific part of technology. Once their privileges were removed, this type of technological capability was no longer appropriate to a competitive environment which rewarded lower costs and quick product updating.

Institutional arrangements and the technological behaviour of enterprises

Institutions as a device of human cooperation have exerted a decisive influence on technological change and economic development. Many authors attribute the success of industrial development in developed countries to the institutions evolved in Western Europe (Rowstow 1975; Hayek 1973; Jones 1988; North and Thomas 1973). The history of technological development in imperial China shows that institutions favourable to rent-seeking through political privilege are the most important factor leading to the stagnation of economic growth. The institutional legacy of traditional Chinese society, such as the state patronage of academic inquiry, the central bureaucracy system and the attitude of bureaucrats to ignore uncodified practical knowledge, still influences the process of technological progress in China.

The empirical findings of this study show that institutional arrangements are of primary importance in affecting the technological behaviour of enterprises. The sample enterprises were under the same institutional arrangements of state ownership, guided by the same policy measures and experienced similar reform initiatives. They behaved similarly in importing and utilizing foreign technology, and lost their market share and reduced their profits in a similar manner when they were faced with the competition from non-state firms which were under different institutional arrangements and behaved differently.

The role of institutional arrangements in affecting the performance of technology transfer by enterprises is not one factor on the same analytical level as other explanatory factors. The importance of institutional arrangements lies in their impact on how firms perceive other factors and in the constraints they face on how firms react to other factors. As pointed out by North (1990: 3) and Pejovich (1990: 4), institutional arrangements guide human interactions and therefore affect transaction costs and transformation costs of human interactions. Therefore, the impact of those factors stressed by many people (see Reddy and Zhao 1990: 299), such as the types of technology, the terms of technology transfer contracts, and the initial technological capabilities of the recipient firms, on the process of technology transfer are constrained and shaped by the institutional arrangements governing the operation of enterprises.

Active learning is vital for enterprises to master the imported technology. However, the active learning efforts of the recipient enterprises are also motivated by the institutional arrangements under which they operate because institutional arrangements determine whether they will benefit from their learning efforts. Under an ambiguous property rights structure, enterprises tend to invest less in learning, because they do not have the exclusive rights to the result of the learning efforts. On the contrary, they tried to divert investment from long term technology development to those activities which could generate quick returns.

The poor performance of the sample enterprises in market competition shows how old institutional arrangements had shaped their expectations and led to certain actions in the past which turned out to be inappropriate to a changed market environment later on. Therefore, in order to understand the impact of institutional arrangements, history, that is the changes in macroeconomic environment of institutional arrangements and market conditions, should be taken into account. This is because the choices of economic agents in the past affects their current actions due to path-dependence in the process of economic change (Dosi and Metcalfe 1991). In the case of state tape recorder enterprises in China, the duplicated imports of tape recorder assembly lines and production facilities by state enterprises reflects the bandwagon effects caused by the erroneous price signals due to the protection of the tape recorder market. Moreover, the capital intensive technology adopted initially was one of the important reasons that the sample state enterprises were unable to compete with non-state enterprises after the late 1980s, because this type of technology was difficult to adapt to the changed environment in which more flexible labour intensive technology offered a competitive advantage.

POLICY IMPLICATIONS

As we discussed in Chapter 6, in China, the government policy towards industry and technology development was mainly for the state sector and the policy instruments were integrated with the governance and control of government over its subordinate SOEs. Therefore, the technological policy in China was mainly implemented through direct intervention into the technological activities of state enterprises, rather than through indirectly changing the economic environment. In the case of the tape recorder industry, due to the institutional arrangements of ambiguous property rights in SOEs, it is not surprising that firms would seek 'rents' from the policy.

In the case of tape recorder production during the mid-1980s, the 'rents' from the government policies of technology imports and of renovating existing enterprises were quite high, because of the import substitution for the consumer durable and subsidized financial resources (see Chapter 4). Consequently, the technological behaviour of SOEs shaped by the institutional arrangements was further biased in the sense that goals of technology imports policy to cut import costs and to improve long term technological capabilities of firms had not been achieved. There were many problems in importing and utilizing foreign technology as showed in the cases of the sample tape recorder firms, such as the duplicated imports of assembly lines and machinery, inefficiencies and delay in installing the imported equipment, the persistently high production costs, and the lack of learning efforts by the recipient state firms.

The example of technology imports by Chinese state tape recorder enterprises shows the necessity of further reforms. In order to regulate the technological behaviour of SOEs, the most important item on the reform agenda of China is to establish a property rights structure for state enterprises. In the past fifteen years, reform of SOEs was mainly implemented through the enlargement of the decision making autonomy of enterprises within the framework of government ownership. However the separation of 'management' and 'ownership' was not as simple as central policy makers had expected. It caused classic common property and agency problems, because of the further blurring of the property rights structure of SOEs under the government ownership hierarchy. Consequently, SOEs did not behave like independent economic entities responsible for their own performance. On the one hand, they tended to increase their near term benefits by sacrificing their long term development. On the other hand, they were still controlled by the government which often wanted to achieve other goals, such as social stability, even at the expense of inefficiency and technology decay in state enterprises.

Therefore, in order to regulate the technological behaviour and encourage the improvement of technological capabilities of SOEs, the reform of state enterprises should further deepen to construct a well delineated property rights structure for SOEs. At present, the issue of how the state, as the owner of state assets, can effectively ensure that the value of state assets can be maintained and increased without the use of bureaucratic power has not been solved yet. However, from the empirical findings of this study, the appropriate policy should include the following considerations.

First of all, the current system of governments at various levels owning and controlling their subordinate SOEs has to be reformed. State enterprises must become independent economic entities, so that the direct intervention of government agencies into the operation of enterprises can be stopped, and the government will no longer bear the unlimited responsibility for enterprises. Second, in order to protect the interests of the state as the owner, appropriate institutional arrangements effectively supervising the operation and management of state assets should be established. The property rights to the state assets should be exclusive and transferable, so that the common property problem can be reduced and the resources of enterprises, including the technological capabilities embodied in capital goods, people and organizations, can be used in their highest valued uses. Specifically, for the large- and medium-sized enterprises, corporatization under a unified framework of corporate law may be an appropriate way which can improve the efficiency and motivate the long term development of SOEs while preventing the value of state assets from being dissipated. For small enterprises, diversification of ownership, such as lease and sale of state assets to non-state firms, may be an appropriate option.

At the current stage of reform, the rigid labour system and the obligations of SOEs to provide social security for their employees have become the major obstacles to the ownership reform in SOEs (Harrold and Lall 1993; Jefferson and Rawski 1994). They restrict labour mobility and are also responsible for the employees' income maximization propensity in SOEs. Because of these obligations, SOEs could not adjust their size of operations to the changing market conditions by adding or shedding their labour force, let alone the re-allocation of their industrial assets in the forms of bankruptcy and merger, which meant that at least a part of the employees of affected firms would lose their jobs. Therefore, in order to transform SOEs to become independent economic entities responding to market signals, market-oriented reforms in the social security system and labour system should be implemented to delink the social functions of SOEs. In the respect of social security system reform, the enterprise-based system should be replaced by insurance systems of unemployment, retirement and health care financed by social security funds. As far as labour system reform is concerned, enterprises should be granted more autonomy in managing their work force to break the lifetime employment in SOEs. Furthermore, enterprise management should be given the authority to determine wage levels. Reforms in social security system and labour system will allow SOEs to be more responsive to market signals and pave the way for the ownership reform in SOEs.

Notes

1 INTRODUCTION

1 See Chapter 3 of this study and Conroy (1992).
2 See Chapter 3 of this study and Sheng *et al.* eds (1991).
3 The term of 'open door' policy is used in China to refer to the change in economic policy after 1978 from inward looking self-reliance to outward orientated development strategy.
4 For the definition of technology in this study, see discussions below.
5 Tape deck is one of the main components of cassette tape recorders (see Chapter 4).

2 THEORETICAL FOUNDATIONS AND RESEARCH METHODOLOGY

1 In some literature, technology is classified as embodied technology and disembodied technology. The former refers to capital goods and the latter includes other forms of technology, such as technical documents and blueprints.
2 For detailed discussions on the CMRS system, TVEs and FDI during the 1980s in China, see Chapter 3.
3 According to the X-efficiency theory of the firm, the effectiveness of the utilization of production inputs depends on the efforts of individual employees. Because the effort is to a certain degree variable, it cannot be assumed that production inputs (including labour) are used as effectively as physically possible. The difference between the maximal effectiveness and the actual result in utilizing inputs is a measure of the degree of X-inefficiency (Leibenstein 1978).

3 ECONOMIC REFORMS AND TECHNOLOGY IMPORTS IN CHINA DURING THE 1980s

1 For discussions of the pre-reform political economy of China, see Riskin (1987), Perry and Wong (1985) and Cheng (1982).
2 In China, people are classified into urban residents and rural residents through a household registration system. Labour mobility was strictly restricted both geographically and from rural status to urban status (see Byrd and Lin 1990: 4).
3 It was stated that 'vitalizing state enterprises is the key to economic reforms in the urban areas' (The Third Plenum of the Eleventh National Congress of the Chinese Communist Party 1978).
4 Separation of 'management' and 'ownership' of state enterprises has been one of the aims of enterprise reform since the late 1970s (Tam 1992: 9). Nevertheless, managers of state enterprises had not obtained substantial autonomy in running

state enterprises. In July 1992, the State Council promulgated 'The Regulations for Transforming the Operating Mechanism of the State-owned Enterprises' which reiterated that management of state enterprises should have autonomous rights in fourteen key areas (Harrold and Lall 1993: 41). Despite this, it has been reported that state enterprises still had not obtained some of the key autonomous rights (Zhu 1993).

5 For discussions on China's industrial reforms in the 1980s, see Tidrick and Chen (1987), Harding (1987), Byrd (1987), World Bank (1990), and Wu Jinglian (1993a).

6 For development of TVEs in the 1980s, see Byrd and Lin eds (1990).

7 For discussions of FDI in China during the reform era, see Pomfret (1991) and Kamath (1990).

8 See Wu Jinglian (1993b) for reasons for the success of price reforms.

9 It was reported that the prices of most commodities including production inputs were no longer controlled by the government. See Renmin Ribao (People's Daily) overseas edition of 2 September 1992. Also see Luo (1991).

10 Chinese enterprises are generally classified as large, medium and small ones, based on production capacity or the value of fixed assets (World Bank 1983: 141).

11 'Eating from a big iron bowl' is a Chinese saying that means the egalitarian distribution of income among employees.

12 In July 1992, the State Council promulgated 'Regulations for Transforming the Operational Mechanism of Industrial Enterprises Owned by the Whole People', which gave state owned enterprises autonomy in fourteen aspects concerning production, marketing, management and investment.

13 Foreign invested firms in China include equity joint ventures, cooperative ventures and wholly foreign owned firms. They are called '*Sanzi Qiye*' which literally means 'three types of foreign invested firms'. They normally enjoy preferential treatment from the Chinese government. For details of foreign direct investment in China, see Pomfret (1991). It has been reported that the share of state sector in total industrial output reduced from nearly 80 per cent in the late 1970s to around 50 per cent in the early 1990s (SBS 1991: 394; Harrold and Lall 1993: 4).

14 In 1978, SOEs' GVIO accounted for 77.6 per cent of the national GVIO. In 1995, it reduced to only 34 per cent (SSB 1996: 403).

15 The total losses made by state enterprises increased by tenfold from RMB ¥3.4 billion in 1980 to RMB ¥34.9 billion in 1990, while their total profits and taxes increased by only 166 per cent during the same period (SSB 1991: 410). It was reported that in 1994, more than 60 per cent of state enterprises suffered losses (Wang 1995: 36).

16 See Chapter 2.

17 The reasons why CMRS cannot define exclusive property rights are discussed in Chapter 2.

18 The administration of Chinese economy was conducted by a 'cellular structure' in which central ministries and their local counterparts supervised the respective functions of enterprises (vertical control), while the local governments (provincial or municipal) owned or controlled enterprises under their jurisdictions (Byrd 1987: 6–9; Donnithorne 1972).

19 This section is based on SEC (1985), MoFERT (1989) and Conroy (1992), and the interview of the author with officials of the State Economic and Trade Commission of China.

20 In 1981, the State Council of China decided to carry out technological renovation for existing enterprises and formulated a series of measures later on to facilitate the renovation programme (see SEC 1985).

21 Localization means the replacement of the imported parts and components of given final products with locally produced ones. See Conroy (1992: 228–30).

22 Shanghai, Tianjin and Dalian were granted the authority to approve projects whose total investment was less than ten million Yuan (changed to 30 million Yuan afterwards). Shanghai and Tianjin were allocated US $300 million and US $200 million respectively from the total budget of US $3,000 million (SEC 1985: 368–81).
23 These preferential forms of treatment were in effect throughout the 1980s and the early 1990s, mainly for those projects which were in the national plan of technological renovation (MoMBEI 1993, Part VI: 4). However, as the enterprise reforms proceeded in the late 1980s, government grants and bank loans were gradually replaced by the enterprises' own funds. In addition, exemption from import tariffs was tightened to halt duplicated imports of foreign equipment (see below).
24 The State Economic Commission was dismantled in 1988. Its function of administering existing enterprise renovation with imported technology was transferred to the State Planning Commission, which in turn returned this responsibility back to the newly established State Economic and Trade Commission (SETC) which was set up in 1992 and is actually the restoration of SEC.
25 The norm for total investment of individual projects was RMB ¥10 million in the first years of the 1980s and increased to RMB ¥30 million later on. In addition, technology import projects of which the total foreign exchange spending larger than US $5 million were also subject to central approval (SEC 1985: 35 and 47).
26 The following discussion is based on MoFERT (1989: 321–46).
27 The norms for machinery and equipment imports for technology renovation projects were US $0.5 million for equipment and US $0.2 million for instruments. In theory, contracts were not allowed to be split into several small contracts to avoid central control. However, localities and enterprises could generally find excuses to do so (interview information).
28 After 1985, the lists of technology and equipment which are under the control of the import licensing system and the import adjustment tax had been updated every year until 1992. In 1992, there were forty-four types of technology and equipment on the import prohibition list. Facing pressure by the American government, the Chinese government is planning to abolish the imports barrier on technology and equipment gradually (MoFERT 1989: 321–46; MoMBEI 1993, Part II: 2–10).
29 For example, tariffs on 'software' technology would be waived entirely if the project was in the national plan of technology renovation, while the tariffs on 'hardware' technology, that is capital goods, would be reduced by only 50 per cent (MoMBEI 1993, Part VI: 3). In addition, in order to encourage enterprises to use disembodied technology, imports of 'software' technology were much easier to be put in the plan and allocated funds (Sheng *et al.* eds 1991: 34).

4 TECHNOLOGY TRANSFER IN THE CHINESE TAPE RECORDER INDUSTRY

1 Statistics of SSB and CAIA show that in 1980, only 1.75 per cent of Chinese families in urban areas owned tape recorders. In 1993, this figure was 76.2 per cent (SSB, various issues).
2 The tariff for tape recorders (including CKD kits) was as high as 200 per cent in 1985 (MoFERT 1989: 332).
3 In 1985, the ratio of profits and taxes to total assets was 24.02 per cent for total industry and 24.67 per cent for the electronics industry, 26.3 per cent and 24.3 per cent lower respectively than that for the tape recorder industry (SSB 1986: 325).
4 Since 1986, market competition has forced decreases in the price of tape recorders.

From 1986 to 1992, the average annual change of sales volume of tape recorders was –0.96 per cent, while the annual change of sales at current price was –2.49 per cent (Table 4.1.). If the high inflation rate during this period and the upgrading of tape recorders are taken into account, the downward change of the real price of tape recorders was dramatic after 1985.

5 According to the regulations issued by the government in 1985, tape decks were on the list of mechanic and electronic goods which were restricted to import. As a result, in addition to the normal tariff which was already as high as 60 to 100 per cent of the value of imported tape decks, an import adjustment tax (equal to the normal tariff) was levied (MoMBEI 1993: Part VI).

6 Township and village enterprises (TVEs) are collectively owned and operated by a township or a village. TVEs are different from urban collective enterprises in financing and administration (see Chapter 4 of this study and Byrd and Lin eds 1990).

7 'Socialist transformation' refers to the nationalization movement in the mid-1950s during which small private firms were transformed into state or collectively owned firms (see Barnett 1981).

8 In China, state firms can be classified into central firms and local firms. The former are usually large enterprises in heavy industry and controlled directly by central industrial ministries while the latter are medium- and small-sized firms under the direct supervision of local governments. However, central industrial ministries are responsible for the development plans and production coordination of local firms. This coordination is carried out via the relevant local industrial bureaux.

9 The party still retains the right to appoint the managing directors of state enterprises. That secures the control of the party over enterprises and constrains the power of managers (Qian Yingyi 1995: 22).

10 See articles on technology imports in the Almanac of China's Foreign Economic Relations and Trade (MoFERT, various issues).

11 For the definition of 'localization', see footnote 21 of Chapter 3.

12 In the practice of China's statistics, GVIO is used to measure the value of industrial output produced by an enterprise. It is similar to turnover of a firm. However, it is at constant prices (SSB 1990: 481).

5 THE TECHNOLOGICAL BEHAVIOUR OF CHINESE SOEs

1 The administrative process is reported in the case studies of WRF in Chapter 6.

2 See Chapters 3 and 7 for the different regulations for SOEs and non-state firms.

3 In the early 1980s, the central planners tried to coordinate the production of certain goods by selecting appointed producers which were given priority for bank loans and other assistance from central industrial ministries.

4 The official exchange rate in 1985 was 2.94 RMB Yuan to 1 US dollar (MoFERT 1990: 838). The official exchange rate of RMB ¥ to US $ was much lower than that in the free market. For example, in 1987, the exchange rate of RMB ¥ to US $ in the free market was 4.7 to 5.7, while the official rate was 3.7 (World Bank 1990: 19). More importantly, foreign exchange was strictly controlled by the state until recently. It was very difficult for enterprises to obtain foreign currency if their investment projects were not in the plan.

5 In 1982, the State Council made a decision to technologically renovate the existing enterprises and adopted many preferential policy measures, including exemption from tariff for imported capital goods and the provision of subsidized bank loans, to encourage state enterprises to carry out these projects (see Chapter 4 and SEC 1985: 7).

6 The central government planned to use US $3 billion to renovate the technology of

existing enterprises in the period from 1983 and 1985. These financial resources were allocated administratively among localities (see Chapter 3).

6 TECHNOLOGY IMPORT PROJECTS OF FOUR SOEs

1 The case study of CRF is based on the interviews with the Chief Engineer of CRF and the Managing Director of WEA. Other information sources include the statistics and technical documents of CRF.
2 In a compensation trade agreement, a foreign firm provides technology, equipment and inputs to a Chinese firm which uses a proportion of its products manufactured with the imported equipment and technology as the payment to the foreign firm (Ho 1990).
3 The reasons for the lack of specialization in China's state enterprises are discussed in Chapter 8.
4 A desk tape recorder is large in size and not portable. It is used as a furnishing and therefore was popular in the early 1980s in China.
5 In late 1985, the central government issued a decree stipulating that assembly lines and CKD kits of tape recorders were no longer to be imported (see Chapter 3).
6 In China, state firms have the obligation to pay for pensions and medical care for their retired employees because there are no social welfare schemes (see Harrold and Lall 1993).
7 The Ministry of Machine Building and Electronic Industry (MoMBEI) consists of two branches covering the machine building industry and the electronics industry. These two parts operate fairly independently and sometimes they are two separate ministries. GBCBI is in charge of broadcasting equipment as well as radios, recorders and television sets. It belongs to the electronics industry branch of MoMBEI.
8 For details of the government policy on technology transfer in the 1980s, see Chapter 3.
9 The production technology of metal tape deck was imported by Wuxian Wireless Factory which is also included in my case studies (see Chapter 6).
10 The localization rate is defined as the percentage of domestically produced parts in terms of the value of total parts.

7 IMPACTS OF MACROECONOMIC FACTORS ON THE TECHNOLOGY TRANSFER PROCESS IN THE CASE FIRMS

1 State firms were not allowed to hire workers with rural residence status.
2 According to the interview with the Vice General Secretary of CAIA, the real profits of Jianghai were probably higher than this figure, because the prices of both imported CKD kits and exported final products were controlled and, thus, might be manipulated by the Hong Kong company. Because the information about the total capital employed by Jianghai was not available, the ROI of Jianghai could not be calculated.
3 The following discussion is based on the interviews with the senior staff member of CAIA and managers of WWF.
4 The tape decks of Jianghai were simpler than those of WWF. If measured in terms of price, two Jianghai's tape decks were roughly equal to one WWF's tape deck.
5 It was reported that the interest rate of the bank loans for CRF's second project was only 5.6 per cent, which was lower than the average annual inflation rate of 7 per cent during the period from 1985 to 1987 (SSB 1991: 229).
6 The documents used include the preliminary project proposal, the project design,

the feasibility study report, and the applications for bank loans, import permission and a foreign exchange quota. The relevant agencies from which WRF had to obtain approval include the Provincial Planning and Economic Commissions, the Provincial Bureau of Electronic Industry, the State Economic Commission, the General Bureau of Broadcasting and Television Industry, the State Import and Export Commission, the Construction Bank, the Bank of China and the Administration of Foreign Exchange.

7 According to SSB (1994: 231), the average annual CPI increase for urban dwellers was 11.1 per cent during the period 1987 to 1993.

8 When profits continued to decrease, however, the increase of employees' income could not be sustained, as is shown in the case of Westlake in Chapter 6. This is because as a result of reform, the retained profits of SOEs was linked with their total profits.

Appendices

APPENDIX 1. THE SAMPLE ENTERPRISES OF THE CHINESE TAPE RECORDER INDUSTRY

A Manufacturers of tape recorders

1 Beijing Wireless Instruments Factory (BWIF), Peony Audio Group, Beijing.
2 Beijing Chunyan Wireless Factory (BCWF), Beiying Group, Beijing.
3 Tianjin Beihai Wireless Factory (TBWF), Tianjin.
4 Tianjin No. 731 Factory (T731), Tianjin.
5 Nanjing No. 3 Recorder Factory (N3RF), Jiangsu.
6 Nanjing Dongfang Wireless Factory (NDWF), Jiangsu.
7 Nanjing Dangqiu Machinery Factory (NDMF), Jiangsu.
8 Nantong Wulin Factory (NWF), Jiangsu.
9 Nantong Recording Equipment Factory (NREF), Jiangsu.
10 Yangcheng Wireless Factory (YWF), Jiangsu.
11 Canzhou Recorder Factory (CRF)*, Jiangsu.
12 Canzhou Audio Equipment Factory (CAEF), Jiangsu.
13 Wujin Electronic Audio (WEA), Jiangsu.
14 Yangzhi Wireless General Factory (YWGF), Jiangsu.
15 Baocheng Wireless Factory (BWF), Jiangsu.
16 Wuxi No. 15 Wireless Factory (W15WF), Jiangsu.
17 Shanghai No. 22 Wireless Factory (S22W) Shanghai.
18 Shanghai No. 13 Wireless factory (S13W), Shanghai.
19 Shanghai Audio Equipment Factory (SAEF), Shanghai.
20 Shanghai No. 742 Factory (SH742), Shanghai.
21 Shanghai Ruijin Audio Factory (SRAF), Shanghai.
22 Hangzhou Westlake Recorder Factory (Westlake)*, Zhejiang.
23 Ningbao Haining Wireless Factory (NHWF), Zhejiang.

B Manufacturers of tape decks for tape recorders

24 Beijing Huabei Recorder Factory (BHRF), Beijing.
25 Tianjin Jinhe Wireless Factory (TJWF), Tianjin.
26 Weihai Recorder Factory (WRF)*, Shandong.
27 Nanjing No. 8434 Factory (N8434), Jiangsu.
28 Canzhou relay Plant (CRP), Jiangsu.
29 Wuxian Wireless Factory (WWF)*, Jiangsu.
30 Shanghai Huamei Wireless Factory (SHWF), Shanghai.
31 Shanghai No. 2532 Factory (S2532), Shanghai.

* The sample firms of case studies.

APPENDIX 2. THE PRODUCTION PROCESSES OF TAPE RECORDERS AND TAPE DECKS IN CHINESE ENTERPRISES

The process of cassette tape recorder production

The production process of cassette tape recorders in a typical Chinese tape recorder producing enterprise can be shown as in Figure A.1. The major production activities include: (1) final product assembling; (2) electric parts soldering and assembling; (3) plastic parts manufacturing; and (4) mould/die making. The task of mould/die making is to make moulds used as implements for extrusion moulding presses to produce plastic parts. The major equipment used in mould making is mainly metal cutting machine tools and heat treatment equipment. Mould/die making is considered to be a process involving a high level of technological capabilities. Not only is the requirement on equipment precision high, but it also needs designing, processing and assembling skills. Plastic parts manufacturing includes press moulding of plastic parts and surface mounting of plastic parts. The equipment for the former is an extrusion moulding press. Surface mounting includes electroplating and silk-screen printing of plastic parts, which are done with a production line consisting of coating and printing machines. The quality of plastic parts is dependent on the quality of the mould and the coating process technique and materials, such as coating powders. The soldering and assembling of electric parts include the soldering of electric circuit boards and the assembling of outside sourced electrical parts and components, such as tape decks, speakers and switchers with the electric circuit board to make up the electric part of a cassette tape recorder. The soldering production line comprises a component connector, wave soldering equipment, soldering point cutter and various electronic measuring instruments to inspect the quality of assembled electrical parts. The final product assembling is done through an assembly line. Along the assembly line, the assembled electric part, outside sourced components and plastic parts are fitted with the outer case of tape recorder. In the last two stages of assembly, a central signal generator is used to adjust and measure the electronic audio quality.

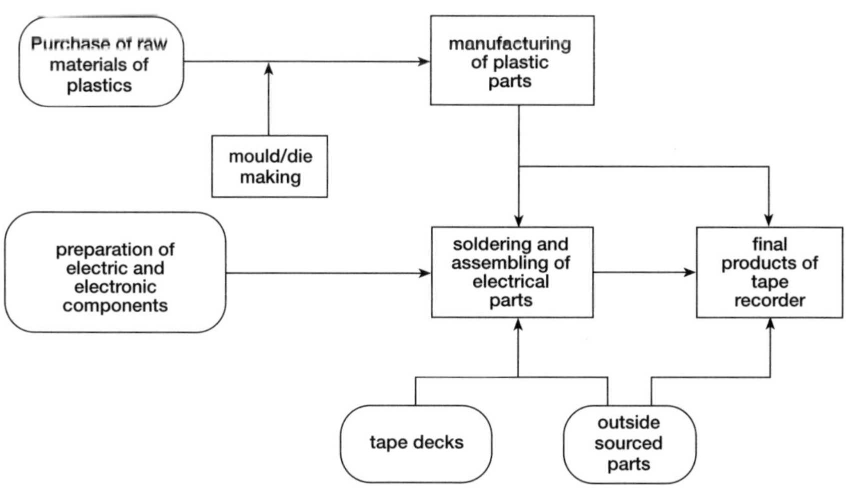

Figure A.1 The process of cassette tape recorder production

B The process of tape deck production

Most tape recorder assemblers do not produce tape decks by themselves, because of the different technology involved and the need for scale economies. All tape deck producers, however, also assemble tape recorders. Figure A.2 illustrates the production process of tape decks in Chinese enterprises. A tape deck consists of metal parts, plastic parts and electrical components. Normally, Chinese tape deck manufacturers produce most of the metal parts and plastic parts by themselves. The metal parts of a tape deck are divided into two groups. The first group includes metal transmission levers, brackets, and spacers, which are made with punches. The second includes spacers, bearings, axles, and flywheels, which are processed by machine tools. The plastic parts of tape deck are push-buttons and face boards, etc. The auxiliary parts include standardized parts and non-metal parts. A few enterprises produce some auxiliary parts, such as nylon spacers and springs, by themselves. The outside sourced components are mainly magnetic heads and mini-motors which are new products to China and produced by other enterprises with imported technology. The assembling of tape decks is done on a production line along which the parts and components are put together by soldering and screwing.

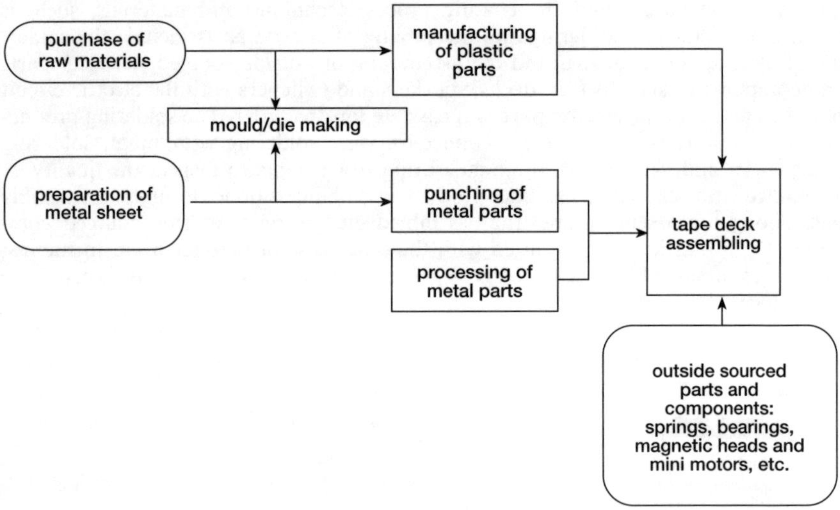

Figure A.2 The process of tape deck production

References

Amendola, M. and Gaffard, J. (1988) *The Innovative Choice*, Oxford: Basil Blackwell.

Arrow, K. (1962) 'Economic Welfare and the Allocation of Resources for Invention', in Nelson ed. 1962: 609–23.

Baert, L., Theunissen, L., and Vergult, G. eds (1988) *Digital Audio and Compact Disc Technology*, Oxford: Heinemann Newnes.

Balazs, E. (1964) *Chinese Civilization and Bureaucracy*, New Haven: Yale University Press.

Barnett, D. (1981) *China's Economy in Global Perspective*, Washington, DC: Brookings Institution.

Barzel, Y. (1989) *Economic Analysis of Property Rights*, Cambridge: Cambridge University Press.

Baumol, W. (1990) 'Entrepreneurship: Productive, Unproductive, and Destructive', *Journal of Political Economy*, 98(5): 893–921.

Bodde, D. (1991) *Chinese Thought, Society and Science*, Honolulu: University of Hawaii Press.

Brewer, J. and Hunter, A. (1989) *Multimethod Research – A Synthesis of Styles*, Newbury Park: Sage Publications.

Buchnan, J. (1980) 'Rent Seeking and Profit Seeking', in Buchnau, Tollison, and Tullock eds 1980: 3–15.

Buchnau, J., Tollison, R., and Tullock, G. eds (1980) *Towards a Theory of the Rent-seeking Society*, College Station: Texas A&M University Press.

Byrd, W. (1987) *The Market Mechanism and Economic Reforms in Chinese Industry*, PhD dissertation at Harvard University, Ann Arbor: UMI.

Byrd, W. and Lin, Q. eds (1990) *China's Rural Industry: Structure, Development, and Reform*, New York: Oxford University Press.

Byrd, W. and Tidrick, G. (1987) 'Factor Allocation and Enterprise Incentives', in Tidrick and Chen eds 1987: 60–102.

CAIA (China Audio Industry Association) (1994) *Zhongguo Dianzi Yingxiang Gongye: 1979–1993 (China's Electronics Audio Industry: 1979–1993),* Shanghai: CAIA, mimeo.

Cao, J. (1986) 'The Present Condition of, and Problems in China's Technology Imports', *Liaowang* (overseas edition), Nos. 18–20.

Cao, J. (1988) 'Technology Transfer to China', *Intertrade*, Nos. 22–5.

Carlsson, B. and Jacobsson, S. (1993) 'Technological Systems and Economic Performance: The Diffusion of Factory Automation in Sweden', in Foray and Freeman eds 1993: 77–94.

Chantramonklasri, N. (1990) 'The Development of Technological and Managerial Capability in the Developing Countries', in Chatterji ed. 1990: 36–50.

Chatterji, M. ed. (1990) *Technology Transfer in the Developing Countries*, London: Macmillan.

Chen, D. (1995) *Chinese Firms between Hierarchy and Market: The Contract Management Responsibility System in China*, New York: St Martin's Press.

Chen, K., Jefferson, G., Rawski, T., Wang, H., and Zheng, Y. (1988) 'Productivity Change in Chinese Industry: 1953–1985', *Journal of Comparative Economics*, 12: 570–91.

Cheng, C. (1982) *China's Economic Development: Growth and Structural Change*, Boulder: Westview Press.

Cheung, S. (1970) 'The Structure of a Contract and the Theory of a Nonexclusive Resource', *Journal of Law and Economics*, 13 (April): 49–70.

Cheung, S. (1974) 'A Theory of Price Control', *Journal of Law and Economics*, 17(1) (April): 53–71.

Cheung, S. (1986) *Will China Go 'Capitalist'?*, London: The Institute for Economic Affairs.

Cimoli, M. and Dosi, G. (1990) 'The Characteristics of Technology and the Development Process: Some Introductory Notes', in Chatterji ed. 1990: 51–71.

Coase, R. H. (1937) 'The Nature of the Firm', *Economica*, 4 (November): 386–505.

Coase, R. H. (1960) 'The Problem of Social Cost', *Journal of Law and Economics*, 3(1) (October): 423–56.

Cohen, W. and Levinthal, D. (1990) 'Absorptive Capacity: A New Perspective on Learning and Innovation', *Administrative Science Quarterly*, No. 35: 128–52.

Conroy, R. (1986) 'China's Entry into World Markets: China's Technology Import Policy', *The Australian Journal of Chinese Affairs*, No. 15 (January): 19–56.

Conroy, R. (1992) *Technological Change in China*, Paris: OECD.

Cusumano, M. and Elenkov, D. (1994) 'Linking International Technology Transfer with Strategy and Management: A Literature Commentary', *Research Policy*, 23: 195–215.

Dahlman, C. J., Ross-Larsen, B., and Westphal, L. E. (1987) 'Managing Technological Development: Lessons from the Newly Industrializing Countries', *World Development*, 15(6): 759–75.

Demsetz, H. (1982) *Economic, Legal, and Political Dimensions of Competition*, Amsterdam: North Holland Publishing Company.

Denzin, N. (1978) *The Research Act (2nd edition)*, New York: McGraw-Hill.

Dollar, D. (1990) 'Economic Reform and Allocative Efficiency in China's State-owned Industry', *Economic Development and Cultural Change*, 39(1): 89–105.

Donnithorne, A. (1972) 'China's Cellular Economy: Some Economic Trends Since The Cultural Revolution', *The China Quarterly*, 52 (October–December): 605–19.

Dosi, G. (1982) 'Technological Paradigms and Technological Trajectory: A Suggested Interpretation of the Determinants and Direction of Technological Change', *Research Policy*, 11: 147–62.

Dosi, G. (1984) *Technical Change and Industrial Transformation*, London: Macmillan.

Dosi, G. (1988) 'The Nature of The Innovative Process', in Dosi, *et al.* eds 1988: 221–38.

Dosi, G. and Metcalfe, S. (1991) 'On Some Notions of Irreversibility in Economics', in Saviotti and Metcalfe eds 1991: 133–59.

Dosi, G. and Orsenigo, L. (1994) 'Macrodynamics and Microfoundations: An Evolutionary Perspective', in Granstrand ed. 1994: 91–124.

Dosi, G., Freeman, C., Nelson, R., Silverberg, G., and Soete, L. eds (1988) *Technical Change and Economic Theory*, London: Pinter.

Eggertsson, T. (1990) *Economic Behaviour and Institutions*, Cambridge: Cambridge University Press.

Elenkov, D. (1991) 'Strategic Behavior, Transaction Cost and Institutional Perspectives of Foreign Business Activity in an East European Country', Working Paper, MIT, Sloan School of Management, Cambridge, as cited in Cusumano and Elenkov 1994: 205.

Eliasson, G. (1994) 'Technology, Economic Competence and the Theory of the Firm – Discussing the Economic Forces Behind Long-term Economic Growth', in Granstrand ed. 1994: 125–64.

Elvin, M. (1973) *The Pattern of The Chinese Past*, Stanford: Stanford University Press.

Enos, J. (1991) *The Creation of Technological Capability in Developing Countries*, London: Pinter.

Enos, J. L. (1989) 'Transfer of Technology', *Asian-Pacific Economic Literature*, 3(1): 3–35.

Enos, J. and Park, W. (1988) *The Adoption and Diffusion of Imported Technology in the Case of Korea*, London: Croom Helm.

Ernst, D. and O'Connor, D. (1989) *Technology and Global Competition*, Paris: OECD.

Evenson, R. and Ranis, G. eds (1990) *Science and Technology*, Boulder: Westview Press.

Foray, D. and Freeman, C. eds (1993) *Technology and The Wealth of Nations*, London: Pinter Publishers.

Freeman, C. (1974) *The Economics of Industrial Innovation*, London: Francis Pinter, 2nd edition, 1982.

Freeman, C. (1987) *Technology and Economic Performance: Lessons from Japan*, London: Pinter Publishers.

Furubotn, E. and Richter, R. (1991) 'The New Institutional Economics: An Overview', in Furubotn and Richter eds 1991.

Furubotn, E. and Richter, R. eds (1991) *The New Institutional Economics*, College Station: Texas A&M University Press.

Gannicott, K. (1986) *On the Economics of Industrial Research and Development in Australia*, PhD thesis, University of New South Wales.

Giersch, H. (1989) *The Ethics of Economic Freedom*, CIS Occasional Paper, No. 24, Sydney: Centre for Independent Studies.

Gomulka, S. (1990) *The Theory of Technological Change and Economic Growth*, London: Routledge.

Granick, D. (1990) *Chinese State Enterprises*, Chicago: The University of Chicago Press.

Granstrand, O. ed. (1994) *Economics of Technology*, Amsterdam: North-Holland.

Grow, R. (1987) 'How Factories Choose Technology', *The China Business Review*, May–June: 35–9.

Harding, H. (1987) *China's Second Revolution: Reform After Mao*, Washington, DC: The Brookings Institution.

Harrold, P. and Lall, R. (1993) *China: Reform and Development in 1992–93*, World Bank Discussion Paper No. 215, Washington, DC: World Bank.

Hayek, F. (1948/1984) 'The Use of Knowledge in Society', in Nishiyama and Leube eds 1984: 211–24.

Hayek, F. (1973) *Law, Legislation and Liberty*, London: Routledge & Kegan Paul.

Hayek, F. (1978/1984) 'Competition as a Discovery Procedure', in Nishiyama and Leube eds 1984: 254–65.

Ho, K. (1990) *Joint Ventures in the People's Republic of China: Can Capitalism and Communism Coexist?*, New York: Praeger.

Ho, S. and Huenemann, R. (1984) *China's Open Door Policy*, Vancouver: University of British Columbia Press.

Huang, Yasheng (1990) 'Web of Interests and Patterns of Behaviour of Chinese Local Economic Bureaucracies and Enterprises During Reforms', *The China Quarterly*, No. 123 (September): 431–58.

IMF (International Monetary Fund) (1991) *China: Economic Reform and Macroeconomic Management*, Washington, DC: IMF.

James, J. (1989a) 'Public Enterprise, Technology and Employment in Developing Countries', in James ed. 1989, Chapter 2: 21–41.

James, J. (1989b) 'Bureaucratic, Engineering and Economic Men: Decision-making for Technology in the United Republic of Tanzania's State-owned Enterprises', in James ed. 1989, Chapter 3: 42–72.
James, J. ed. (1989) *The Technological Behaviour of Public Enterprises in Developing Countries*, London: Routledge.
Jefferson, G. and Rawski, T. (1994) 'Enterprise Reform in Chinese Industry', *Journal of Economic Perspectives*, 8(2) Spring: 47–70.
Jefferson, G., Rawski, T., and Zheng, Y. (1992) 'Growth, Efficiency, and Convergence in China's State and Collective Industry', *Economic Development and Cultural Change*, 40(2): 239–66.
Jensen, M. and Meckling, W. (1976) 'Theory of the Firm: Managerial Behavior, Agency Costs and Ownership Structure', *Journal of Financial Economics*, No. 3 (October): 305–60.
Jensen, M. and Meckling, W. (1979) 'Rights and Production Functions: An Application to Labor-managed Firms and Codetermination', *Journal of Business*, 52(4): 469–506.
Jiang, Q. and Shi, L. (1993) 'Jishu Chuangxin: Guoyong Dazhongxing Qiye De Xian Zhuang Yu Wenti (Technological Innovation: The Current Situation and Problems of Large and Medium-sized State Enterprises)', *Zhongguo Gongye Jingji Yanjiu (China Industrial Economics Study),* No. 9: 39–42.
Jones, E. (1987) *The European Miracle: Environments, Economies, and Geopolitics in the History of Europe and Asia*, Cambridge: Cambridge University Press.
Jones, E. (1988) *Growth Recurring*, Oxford: Clarendon Press.
Kamath, S. (1990) 'Foreign Direct Investment in a Centrally Planned Developing Economy: The Chinese Case', *Economic Development and Cultural Change*, 39(1): 106–30.
Katz, J. (1985) 'Domestic Technological Innovations and Dynamic Comparative Advantages: Further Reflections on A Comparative Case-Study Program', in Rosenberg and Frischtak eds 1985: 127–66.
Kim, H. (1986) *The Impact of Indigenous R&D and Technology Transfer on Productivity Growth in a Newly Industrializing Economy: The Case of Korea*, PhD dissertation, University of Pittsburgh.
Kirzner, I. M. (1973) *Competition and Entrepreneurship*, Chicago: The University of Chicago Press.
Kirzner, I. M. (1978) 'The Perils of Regulation: A Market-process Approach' in Kirzner 1985: 119–49.
Kirzner, I. M. (1979) *Perception, Opportunity, and Profit*, Chicago: The University of Chicago Press.
Kirzner, I. M. (1985) *Discovery and The Capitalist Process*, Chicago: The University of Chicago Press.
Kirzner, I. M. (1992) *The Meaning of Market Process*, London: Routledge.
Klein, B. H. (1977) *Dynamic Economics*, Cambridge, MA: Harvard University Press.
Kong, R. and Yang, Y. (1991) 'Xing Zongguo Jishu Yingjin Sishi Nian Shuping (Comments on Technology Transfer of 40 Years in New China)', *Guanli Shijie (Management World),* 6: 168–74.
Lall, S. (1987) *Learning to Industrialize*, London: Macmillan.
Langlois, R. ed. (1986) *Economics as a Process: Essays in the New Institutional Economics*, Cambridge: Cambridge University Press.
Langlois, R. and Robertson, P. (1995) *Firms, Markets and Economic Change: A Dynamic Theory of Business Institutions*, London and New York: Routledge.
Lee, K. (1991) *Chinese Firms and the State in Transition: Property Rights and Agency Problems in the Reform Era*, New York: M. E. Sharpe, Inc.
Leibenstein, H. (1976) *Beyond Economic Man*, Cambridge, MA: Harvard University Press.

Leibenstein, H. (1978) *General X-efficiency Theory and Economic Development*, New York: Oxford University Press.

Leibenstein, H. (1980) 'X-efficiency Theory and the Analysis of State Enterprises', paper prepared for the Boston Area Public Enterprise Group. As cited by James, 1989: 5–6.

Leydesdorff and Besselaar, eds (1994) *Evolutionary Economics and Chaos Theory*, London: Pinter Publishers.

Link, A. (1987) *Technological Change and Productivity Growth*, New York: Harwood Academic Publishers.

Liu, C. (1993) 'The Incentive Structures of China's Enterprises Under Contract Responsibility System', *The Journal of Contemporary China*, 2(1): 69–81.

Liu, S. (1993) 'Zhengfu Yu Guiyong Qiye Shouru Fenpei Guanxi De Lilun Sikao (Theoretical Considerations on the Distribution Relationship Between the Government and State Enterprises)', *Zhongguo Gongye Jingji Yanjiu (China Industrial Economics Study)*, No. 7: 12–19.

Loveridge, R. and Pitt, M. eds (1990) *The Strategic Management of Technological Innovation*, Chichester: John Wiley & Sons.

Lu, D. (1991) 'Zailun Zhuanhuan Jingying Jizhi Shi Gaohao Guoyin Dazhongxing Qiye de Guanjian (Transforming the Operation Mechanism is the Key to Improve the Performance of Large and Medium-sized State Enterprises)', *Renmin Ribao (The People's Daily)*, overseas edition, 1 November: 5.

Lundvall, B. (1993) 'User-Producer Relationships, National Systems of Innovation and Internationalisation', in Foray and Freeman eds 1993: 277–300.

Luo, Z. (1991) 'Renzhen Zongjie Shijian Jingyan, Wentuo Tujin Jiage Gaige (Seriously Summarising Practical Experience and Steadily Promoting Price Reform)', *Renmin Ribao (The People's Daily)*, overseas edition, 30 October: 5.

McGuinness, N., Campbell, N., and Leontiades, J. (1991) 'Selling Machinery to China: Chinese Perceptions of Strategies and Relationships', *Journal of International Business Studies*, No. 2: 187–207.

Mansfield, E., Schwartz, M., and Wagner, S. (1981) 'Imitation Costs and Patents: An Empirical Study', *Economic Journal*, 91: 907–18.

Mayer, T. (1993) *Truth Versus Precision in Economics*, Aldershot: Edward Elgar.

Menshikov, S. (1994) 'The Role of State Enterprises in the Transition from Command to Market Economies', *Journal of Evolutionary Economics*, No. 4: 289–325.

MoFERT (The Ministry of Foreign Economic Relations and Trade) (1989) *Liyong Waizi Wenjian Huibian (Compilation of Documents on Utilisation of Foreign Capital)*, Beijing: China Economic Press.

MoFERT (The Ministry of Foreign Economic Relations and Trade), *Almanac of Foreign Economic Relations and Trade of China*, various issues, Beijing: China Financial and Economic Press.

MoMBEI (The Ministry of Machine Building and Electronics Industry) (1993) *Yearbook of Electronics Industry of China 1993*, Beijing: China Electronics Industry Press.

Mowery, D. and Rosenberg, N. (1989) *Technology and the Pursuit of Economic Growth*, Cambridge: Cambridge University Press.

Mytelka, L. (1985) 'Stimulating Effective Technology Transfer: The Case of Textiles in Africa', in Rosenberg and Frischtak eds 1985: 77–126.

Needham, Joseph (1954–1988) *Sciences and Civilization in China*, Cambridge: Cambridge University Press.

Needham, Joseph (1964) 'Science and China's Influence on the World', in Needham 1969: 123–53.

Needham, Joseph (1969) *The Grand Titration*, London: George Allen & Unwin.

Nelson, R. (1959) 'The Economics of Invention: A Survey of Literature', *Journal of Business*, 32: 101–27.

Nelson, R. ed. (1962) *The Rate and Direction of Inventive Activity*, Princeton: Princeton University Press.
Nelson, R. (1987) *Understanding Technical Change as an Evolutionary Process*, Amsterdam: North-Holland.
Nelson, R. (1990a) 'On Technological Capabilities and their Acquisition', in Evenson and Ranis eds 1990: 71–80.
Nelson, R. (1990b) 'Capitalism as an Engine of Progress', *Research Policy*, No. 19: 193–214.
Nelson, R. (1994) 'Economic Growth via the Coevolution of Technology and Institutions', in Leydesdorff and Besselaar eds 1994: 21–32.
Nelson, R. and Winter, S. (1982) *An Evolutionary Theory of Economic Change*, Cambridge, MA: Belknap Press of Harvard University Press.
Nishiyama, C. and Leube, K. eds (1984) *The Essence of Hayek*, Stanford: Hoover Institution Press.
Nolan, P. and Dong, F. eds (1990) *The Chinese Economy and its Future*, Cambridge: Polity Press.
North, D. (1981) *Structures and Change in Economic History*, New York: W. W. Norton.
North, D. (1990) *Institutions, Institutional Change and Economic Performance*, Cambridge: Cambridge University Press.
North, D. and Thomas, R. (1973) *The Rise of The Western World*, Cambridge: Cambridge University Press.
Orleans, L. D. ed. (1980) *Sciences in Comtemporary China*, Stanford: Stanford University Press.
Patton, M. (1990) *Qualitative Evaluation and Research Methods*, London: Sage Publications.
Pavitt, K. (1985) 'Technology Transfer among the Industrially Advanced Countries: An Overview', in Rosenberg and Frischtak eds 1985: 3–23.
Pejovich, S. (1990) *The Economics of Property Rights: Towards a Theory of Comparative Systems*, Boston: Kluwer Academic Publishers.
Perkins, D. (1988) 'Reforming China's Economic System', *Journal of Economic Literature*, 26 (June): 601–45.
Perkins, D. (1994) 'Completing China's Move to the Market', *Journal of Economic Perspectives*, 8(2): 23–46.
Perry, E. and Wong, C. eds (1985) *The Political Economy of Reform in Post-Mao China*, Cambridge, MA: Harvard University Press.
Pomfret, R. (1991) *Investing in China: Ten Years of the 'Open Door' Policy*, New York: Harvester.
Qian, Yingyi (1995) 'Qiye de Zhili Jiegou Gaige he Rongzi Jiegou Gaige (Reform of Governance and Financial Structure of Enterprises)', *Jingji Yanjiu (Economic Research Journal),* No. 1 (January): 20–9.
Ranis, G. (1990) 'Science and Technology Policy: Lessons from Japan and the East Asian NICs', in Evenson and Ranis eds 1990: 157–78.
Reddy, N. and Zhao, L. (1990) 'International Technology Transfer: A Review', *Research Policy*, 19: 285–307.
Riskin, C. (1987) *China's Political Economy: The Quest for Development Since 1949*, New York: Oxford University Press.
Rosegger, G. (1980) *The Economics of Production and Innovation*, Oxford: Pergamon Press.
Rosenberg, N. (1976) *Perspectives on Technology*, Cambridge: Cambridge University Press.
Rosenberg, N. and Birdzell, L. (1986) *How The West Grew Rich: The Economic Transformation of the Industrial World*, New York: Basic Books.
Rosenberg, N. and Frischtak, C. eds (1985) *International Technology Transfer*, New York: Praeger.

Rostow, W. (1975) *How It All Began*, New York: McGraw-Hill.

Sanders, W. and Pinhey, T. (1983) *The Conduct of Social Research*, New York: CBS College Publishing.

Saren, M. (1990) 'Determinants, Processes and Strategies of Technological Innovation: Towards an Interactive Paradigm', in Loveridge and Pitt eds 1990: 205–23.

Saviotti, P. and Metcalfe, S. eds (1991) *Evolutionary Theories of Economic and Technological Change*, Melbourne: Harwood Academic Publishers.

Schumpeter, J. (1934) *The Theory of Economic Development*, Fifth Printing 1967, New York: Oxford University Press.

SEC (The State Economic Commission) (1985) *Jishu Gaizao Jishu Yinjin (Technology Renovation and Technology Introduction)*, Beijing: China Economic Press.

Sheng Shuren *et al.* eds (1991) *Zhongguo Yinjin Jishu Gaizao Xianyou Qiye Shinian Jian (Ten Years of Technology Imports and Existing Enterprise Renovation in China)*, Beijing: China Economic Press.

Simon, H. (1957) *Models of Man*, New York: Wiley.

Sivin, Nathan (1980) 'Science in China's Past', in Orleans ed. 1980: 1–29.

Solow, R. (1985) 'Economic History and Economics', *American Economic Review*, 75 (May): 328–31.

SPC (The State Planning Commission) (1990) 'Quanguo Yinjin Jishu Gaizao Xianyou Qiye Gongzuo Zongjie (The Summary of the Program of Renovating Existing Enterprises with Imported Technology in China)' in Sheng Shuren *et al.* eds 1991: 81–90.

SSB (The State Statistics Bureau of China) *Statistical Yearbook of China*, various issues, Beijing: China Statistical Publishing House.

Stewart, F. (1990) 'Technology Transfer for Development', in Evenson and Ranis eds 1990: 301–24.

Stoneman, P. and Diederen, P. (1994) 'Technology Diffusion and Public Policy', *The Economic Journal*, 104 (July): 918–30.

Summers. L. (1991) 'The Scientific Illusion in Empirical Macroeconomics', *Scandinavian Journal of Economics*, 93 (March): 129–48.

Tam, O. (1992) *State Enterprise Performance and China's Fiscal Reform*, Economics & Management Working Paper, No. 1/1992, The Department of Economics and Management, University College, The University of New South Wales.

TEI (Technical Economics Institute, State Planning Commission) (1988) *San Qian Xiang Yingjin Jishu Xiaoyi Fengxi (The Analysis of Economic Benefits of the 3000 Technology Import Projects)*. Beijing: TEI, SPC, mimeo.

Tidrick, G. and Chen, J. eds (1987) *China's Industrial Reform*, New York: Oxford University Press.

Tornatzky, L. and Fleischer, M. *et al.* (1990) *The Processes of Technological Innovation*, Lexington, MA: Lexington Books.

Tung, R. (1982) *Chinese Industrial Society after Mao*, Lexington, Mass: Lexington Books.

UNCTC (United Nations Centre on Transnational Corporations) (1987) *Transnational Corporations and Technology Transfer: Effects and Policy Issues,* New York: UNCTC.

UNCTC (United Nations Centre on Transnational Corporations) (1988) *Foreign Direct Investment in The People's Republic of China*, New York: United Nations.

Volti, Rudi (1982) *Technology, Politics, and Society in China*, Boulder: Westview Press.

Wade, R. (1990) *Governing the Market*, Princeton: Princeton University Press.

Walder, A. (1989) 'Factory and Manager in an Era of Reform', *The China Quarterly,* No. 118 (June): 242–64.

Wang, D. (1995) 'Shenhua Guoyou Qiye Gaige Xuyao Zhengce Chuangxing (Policy Innovation is Needed to Deepen State Enterprise Reform)', *Zhongguo Gaige (China Reform)*, No. 1: 34–41.

Wang, J. (1990) 'The Investment System', in Nolan and Dong eds 1990: 227–39.

Wang, Y-F. (1993) *China's Science and Technology Policy: 1949–1989*, Aldershot: Avebury.

Wells, L. (1975) 'Economic Man and Engineering Man: Choice of Technology in a Low-wage Country', in Timmer *et al.* eds *The Choice of Technology in Developing Countries,* as cited by James ed. 1989: 45–6.

Westphal, L., Kim, L., and Dahlman, C. (1985) 'Reflections on the Republic of Korea's Acquisition of Technological Capability', in Rosenberg and Frischtak eds 1985: 167–221.

Westphal, T., Kritayakirana, K., Petchsuwan, K., Sutabutr, H., and Yuthavong, Y. (1990) 'The Development of Technological Capability in Manufacturing: A Macroscopic Approach to Policy Research', in Evenson and Ranis eds 1990: 81–134.

Williams, D. (1979) 'National Planning and the Choice of Technology: The Case of Textiles in Tanzania', in Kim *et al.* eds *Papers on The Political Economy of Tanzania*, as cited by James ed. 1989: 46–8.

Williamson, O. (1981) 'The Economics of Organisation: The Transaction Cost Approach', *American Journal of Sociology*, 87(3).

Williamson, O. (1985) *The Economic Institutions of Capitalism: Firms, Markets, Relational Contracting*, New York: The Free Press.

World Bank (1983) *China: Socialist Economic Development*, Washington, DC: World Bank.

World Bank (1990) *China: Macroeconomic Stability and Industrial Growth under Decentralized Socialism*, Washington, DC: World Bank.

World Bank (1991) *World Development Report 1991: The Challenge of Development*, New York: Oxford University Press.

Wu, J. (1993a) *Dazhongxin Qiye Gaige: Jianli Xiandai Qiye Zhidu (The Reform of Large- and Medium-sized Enterprises: Establishing a Modern Enterprises System),* Tianjin, China: Tianjin Peoples Press.

Wu, J. (1993b) 'Jinyibu Shenhua Gaige de Sanda Renwu (The Three Major Tasks in Deepening Reforms)', *Zhongguo Gongye Jingji Yanjiu (China Industrial Economic Study*), No. 6, 1993: 17–19.

Wu, S. (1991) 'Dui Guoying Dazhongxing Qiye De Xianzhuang Yaozuo Kexue Fenxi (The Current Situation of Large- and Medium-sized State Enterprises must be Analysed Scientifically)', *Renmin Ribao (People's Daily)*, overseas edition, 28 October, 1991: 5.

Xie, D. (1993) 'Guanyu Guoyou Daqiye Pochan Ruogan Wenti De Tantao, (Discussions about the Problems in the Bankruptcy of Large State Enterprises)', *Zhongguo Gongye Jingji Yanjiu (China Industrial Economics Study)*, No. 9: 33–8.

Xu, J. and Chen, G. (1991) 'Cong Qiye Jinying Chengbaozhi De Shijian Kan Woguo Qiye Gaige De Chulu (Search for the Solution of Enterprise Reform from the Practice of the Enterprise Management Contract Responsibility System)', *Shehui Kexue Jikan (Compilation of Social Sciences)*, 2: 12–18.

Xu, L. and Fan, D. (1982) *Science and Socialist Construction in China*, New York: Sharpe.

Zhang,W. (1995) 'Gongyouzhi zhong de Weituoren – Dailiren Guanxi: Lilun Fenxi he Zhengce Hanyi (The Principal–Agent Relationship in the State Owned Economy: The Theoretical Analysis and Its Policy Implications)', *Jingji Yanjiu (Economic Research)*, No. 4: 10–20.

Zhou, S. (1987) 'Shilun Dazhongxing Qiye (Preliminary Research on How to Improve Large and Medium-sized Enterprises)', *Guanli Shijie (Management World)*, 1: 14–32.

Zhu, R. (1987) 'Shenru Kaizhan Yinjin Jishu de Xiaohua Xishao, Dali Tuijin Xianyou Qiye de Jishu Jinbu (Deepening the Digestion and Absorption of Imported

Technology, and Promoting Technology Progress in Existing Enterprises)', in Sheng Shuren *et al.* eds 1991: 76–80.

Zhu, T. (1993) 'Tiaoli Banbu Shige Yue (Ten Months after the Promulgation of the Regulations)', *Liaowan (Perspective)*, No. 21: 7–10.

Index